MORE PRAISE FOR *Chasing Kangaroos*

"*Chasing Kangaroos* is almos
mative years as a palaeontol
landscape, and partly a stud
sounds like three reasons nc
absorbing, funny, and wond

"A warmhearted book by an expert and an enthusiast for his subject . . . Mr. Flannery's fervor for the animals of the fifth continent has kept him curious for three decades." —Ruth Graham, *New York Sun*

"Flannery writes with enthusiasm about the natural history of his subject and explorations in his native land . . . and his stories of field research unfold with wit and flair." —Lisa Palmer, *Providence Journal*

"Written with both earthy humor and scientific precision, this book is almost as unique as its subject. . . . This delightful journey of discovery will appeal to fans of Bill Bryson and Mark Kurlansky." —*Booklist*

"Satisfying . . . Tagging along [with Flannery] is always informative, often fun, and frequently thought-provoking." —Brian Alexander, *San Diego Union-Tribune*

"Flannery is a playful Antipodean Stephen Jay Gould and *Chasing Kangaroos* an intellectually infectious recruiting pamphlet for the next generation of ecologists and palaeontologists." —*Sydney Morning Herald*

"Flannery has been fascinated by kangaroos ever since he abandoned prospects for a humanities degree in favor of work for a noted paleontologist and a cross-country summer trip on a sturdy motorbike some thirty years ago. Not only did these untamed youthful adventures excite his interest in further fieldwork, they also provide the frame here for lively chapters filled with colorful Australian characters and occasionally perilous encounters with the continent's scattered Aborigine population. . . . [*Chasing Kangaroos* is] fired by a boundless exuberance that leaps off the page. —*Kirkus Reviews*

CHASING KANGAROOS

CHASING KANGAROOS

A Continent, a Scientist, and a Search
for the World's Most Extraordinary Creature

TIM FLANNERY

Grove Press
New York

First published in 2004 by The Text Publishing Company, Australia

FIRST AMERICAN EDITION

ILLUSTRATIONS

Grateful acknowledgment is made to the following for permission to reproduce the illustrative material:

COLOR PLATES

Plates 1,4 and 12 are from the author's collection; plates 2,3,6,7,8,9,10 and 11 are courtesy of Dave Watts, Dave Watts Photography; plate 5 courtesy of Lochman Transparencies.

BLACK AND WHITE

Special thanks to Peter Schouten for his drawings on pages 158, 161 and 164, and for permission to reproduce the picture on p. 58. Stubbs' kangaroo on p. 20 is reproduced with permission from the British Library; p.98 and p. 110 courtesy of Cindy Hann; p. 156 courtesy of Museum of Victoria; p.222 from the author's collection; p.247 photographer Barry Wilson, reproduced courtesy of the Australian Wildlife Conservancy.

Library of Congress Cataloging-in-Publication Data

Flannery, Tim F. (Tim Fridtjof), 1956–
    (Country)
    Chasing kangaroos : a continent, a scientist & a search for the world's most extraordinary creature / Tim Flannery.-1 st Grove Press ed.
    p. cm.
Originally published: Country: a continent, a scientist and a kangaroo. Melbourne : Text Pub., 2004.
    ISBN-10: 0-8021-4371-7
    ISBN-13: 978-0-8021-4371-6
    1. Kangaroos-Evolution. 2. Macropodidae-Australia. 3. Extinct animals-Australia. 4. Extinction (Biology)-Australia. 5. Natural history-Australia. 6. Human ecology-Australia. 7. Australia-Description and travel. 8. Flannery, Tim F. (Tim Fridtjof), 1956- I. title.
QL737.M35F56 2007
599.2'22-dc22   2006052628

Maps by Tony Fankhauser

Grove Press
an imprint of Grove/Atlantic, Inc. 841 Broadway
NewYork, NY 10003
Distributed by Publishers Group West
www.groveatlantic.com

08 09 10 11   10 9 8 7 6 5 4 3 2 1

To my mother
Valda Joyce Flannery
a woman of boundless compassion and understanding

Contents

Map   x–xi

Introduction: Vanished Country   1

1   A Failed Circumnavigation   7

2   Captain Cook's Kangaroo   15

3   Quokkas, Euros and Stinkers   27

4   The Last of the Frontier   38

5   Of Nailtails and Nailed Tyres   48

6   Kangaroo Essence   56

7   Dead-end in the Inland Sea   70

8   The Mystery of Hopping   79

9   The Brightest Place on Earth   90

10 The Oldest Kangaroo   100

11 Skeletons in the Dead Centre   107

12 Where the Great Roos Came From   115

13 The Age of Kangaroos   124

14 Advancing with Feet or Stomach?   130

15 Grass for the Kangaroos   136

16 Not Formed for Such Work   143

17 Land of Giants   151

18 Is the Answer 46?   167

19 World Conquest   181

20 A Dingo-driven Revolution   188

21 The Age of Mammals in Australia   194

22	The Groote Eylandt	208
23	The True Experts	216
24	Symbols of the New Land	227
25	Oolacunta!	234
26	Re-making Country	241
	Postscript	249
	Family Tree	250–51
	Acknowledgments	252
	General Bibliography	253
	Index	256

'If you want to study the history of this country,
you'll have to have the will to fail.'
Tom Rich

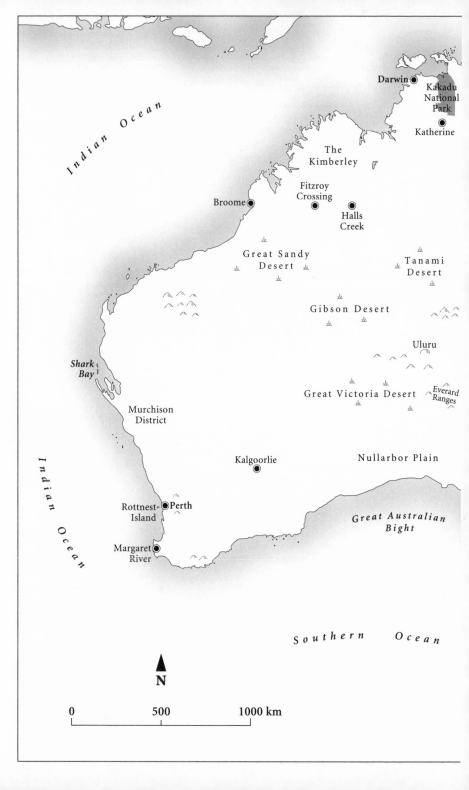

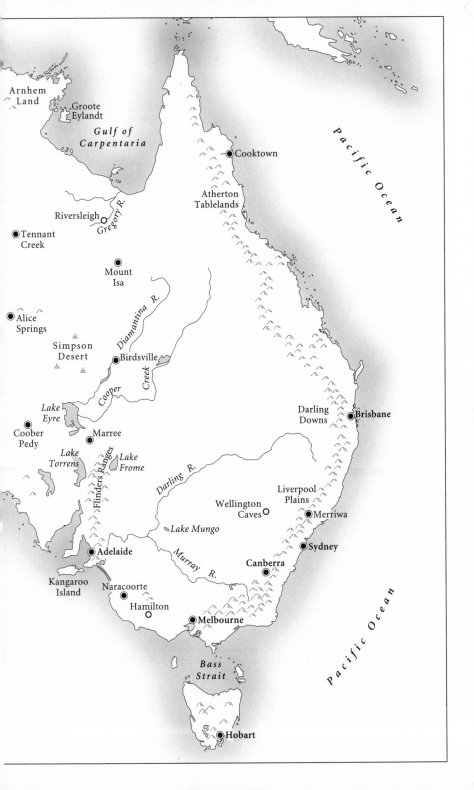

Introduction

Vanished Country

When I was young I met a man whose arse bore the bite-mark of a Tasmanian tiger. David Fleay was one of Australia's most respected naturalists, and he'd received his punctures while bending over to film the creature as it paced in its cage in a Hobart zoo. In my youthful imaginings that scar was the supreme stamp of Australian identity, a badge of honour that lay forever beyond my reach. That was because my eyes opened on the world in Melbourne nineteen years, three months and three weeks after the last tiger closed hers forever. My birthplace was a grand, European-style city of rumbling trams, and men in coats trudging plane-tree-lined streets past Victorian bluestone edifices. I dreamed of finding a thylacine, but by the time I was old enough to travel, even kangaroos and bandicoots had vanished from around Melbourne. So I was a rebellious young man—too angry to take a good look around me—who did not know my country.

Then again, how do you ever know your country? Had I not rejected what I'd been taught at school I might have remembered the sage words:

'by their fruits you shall know them', and perhaps even recollected the *First Fruits of Australian Poetry*, a work that in 1819 eulogised the kangaroo, hailing it as

> thou spirit of Australia,
> that redeems from utter failure...
> this fifth part of the Earth.

The author of that first-published volume of Australian poetry was Justice Barron Field of the Supreme Court of New South Wales. Perhaps only one whose days were spent judging colonial miscreants could have rhymed 'Australia' with 'failure'. Then again, I've always suspected that Australians are a self-deprecating people who, despite their penchant for scattering 'Great' reefs, ranges and bights across the map, were ashamed of their country. Even our national symbol, the kangaroo, has been something of an embarrassment. Many people convert their national animal into *haute cuisine*—venison, wild boar or buffalo steak—but kangaroos are mostly fed to dogs.

The origins of the modern kangaroo industry illustrate how very low in the esteem of Australians the creature had fallen. It began in the 1960s when rabbit shooters, whose game had been ruined by myxomatosis, were forced to search for an alternative. Australians had long welcomed the flesh of the pestilential rabbit to their tables, but they would not touch the kangaroo. So the creatures were shot for skins and the pet-food industry in numbers—nearly nine million in 1966–67—sufficient to threaten them with extermination. The impact of the slaughter prompted naturalist Vincent Serventy to lament that 'from the average person's point of view the kangaroo is now extinct. Nowhere can the average person see a live kangaroo within convenient distance from urban areas.' In 1971, at the first meeting called to discuss the situation, the eminent zoologist Ronald Strahan captured the prevailing sentiment: 'I believe that kangaroo meat is only of commercial interest because it is cheap. The moment we...put effort into its husbandry, we shall find that the game is not worth the candle.'

It astonishes me that such a wondrous creature as the 'spirit of Australia' could plummet so low in the nation's affections, for our first published poet did not miss the mark in his eulogy. So breathtakingly different is the kangaroo that if it did not exist we'd be unable to imagine it: hopping being as marked a departure from running as the orbital engine is from piston and crankshaft, and every bit as efficient. Perhaps familiarity has bred contempt, for unlike the thylacine, whose extinction has endowed it with mythic status, large kangaroos are today so commonplace that most Australians have long ceased to wonder at them. Some even regard them as pests.

Yet they are, in my opinion, the most remarkable animals that ever lived, and the truest expression of my country—not because they appear on everything from the coat of arms to the national airline, but because they have been made by Australia. They are, in short, the continent's most successful evolutionary product. Forged over eons by Australia's distinctive environment, what was originally a tiny possum-like creature has endured a million genetic changes to become a kangaroo. In reading the animal's history we should be able to discover, in distilled form, the story of our country.

In all of life's tenure on Earth no other large creature has achieved the triumph of hopping. True, some rodents hop, but even the largest rodent hopper—South Africa's springhare—is twenty times smaller than a grey kangaroo. Humans are proud of walking upright, but that is an accomplishment shared with many creatures. Hopping is an accomplishment more akin to the development of human language, for as with speaking it is a singular evolutionary achievement.

Yet hopping is only one aspect of a revolutionary design that has made the large kangaroos the most successful of Australia's marsupials. In an age when so many of their relatives have become extinct they are true survivors, for despite centuries of hunting, persecution and competition from supposedly superior herbivores such as cattle, horses and sheep, today they are more abundant than ever. Aerial surveys reveal that there are at least 57 million individuals of the four largest species (red, the euro, and the eastern and western grey). But the success of kangaroos must also be measured by another yardstick— the diversity and breadth of adaptation of the family as a whole.

You might imagine that hopping is such a specialised trait that it has constrained their evolutionary development; after all, you cannot hop backwards. Yet the seventy-odd species of kangaroo, wallaby and rat-kangaroo that currently make up the kangaroo family (we are still counting because scientists keep discovering new ones) have made their homes in a staggering variety of habitats. At least ten species of tree-kangaroo inhabit the treetops of tropical rainforests where they eat fruit and leaves and live like monkeys do on other continents. Some rat-kangaroos excavate burrows under the Australian deserts where they pursue a rabbit-like existence (though their burrows are deeper and more complex), while others tough it out on the surface of some of the most hostile, arid and unpredictable wastelands of our planet. It is no exaggeration to say that every square kilometre of the enormous region stretching from Indonesia's Wallace's Line to Tasmania is, or was, occupied by at least one member of the kangaroo family. They are the chief herbivores of this expansive realm, and in the sheer exuberance of their evolutionary branchings they far outstrip the great mammalian success stories of other continents such as horses, deer or antelope. Yet so uninteresting did these amazing creatures seem to most Australians that until 1963 no one had thought to ask how kangaroos hopped, or why. As late as 1970 no one knew that the grey kangaroos seen in paddocks from Perth to Cooktown were two distinct species— a discovery as startling as if scientists suddenly realised that two species of red deer roamed Britain, or that two species of bison lived on America's Great Plains. And as late as 1980 no one had any idea about the early evolution of the family, for no fossils older than a few million years had been studied. The year 1995 brought the revelation that a black-and-white kangaroo resembling a small panda lives atop the highest mountains of Indonesia. And today the discoveries continue: in 2003 a banded hare wallaby from South Australia was named—the only surviving specimen having lain unrecognised in a European museum for 150 years, while at least two New Guinean tree-kangaroos are yet to be classified and receive their scientific binomen. So rich is the seam tapped by researchers of kangaroo biology that fundamental discoveries continue to emerge at a rapid rate, indicating that, despite all we have learned, we are still at the beginning of understanding these animals and the country that shaped them.

Although they are diverse, the living kangaroos are a mere shadow of the stupendous variety that once existed, for as recently as 50,000 years ago there were over 100 kangaroo species, including some true giants. I first became aware of these ancient behemoths in the early 1970s when, working as a teenage volunteer at the Museum of Victoria, I was entrusted with cleaning the fossilised skeletons of extinct kangaroos that were twice the size of any living species. My studies later broadened to become a quest in time and space aimed at understanding the entire evolution of the kangaroos, from their obscure beginnings to the monsters of the ice age—great flesh-eating kangaroos and kangaroos with ape-like faces—that seemed to have stepped out of a Grimm fairytale.

I've come to think of my work as piecing together an ever-changing jigsaw in which my fellow Australians—animals, plants and people— are all parts. I soon realised though, that most of the puzzle's pieces remained undiscovered, and so I've searched for them in Australia's deep history, wherein lie buried answers to fundamental questions such as where kangaroos come from, and why they hop. The fossils that shed light on these matters are often found in the remote outback, where floods fill Lake Eyre and droughts blight rangelands as extensive as western Europe. It's a wondrous land—full of surprises and subtle beauty. Because the puzzle is still incomplete, the story I'm about to tell makes some leaps. But stick with me for the ride, even if it seems bumpy at times, for I'm sure you'll find the journey worthwhile.

The story begins in Melbourne in 1975, when I could, had I got off my unpunctured arse and roamed the bush instead of dreaming about thylacines, have discovered my very own kangaroo species—right there in Victoria. This is no small matter, for ten million years of evolution have given us only seventy living species; so to claim one as your own is quite an achievement. Neil Armstrong might have walked on the moon by 1975, but East Gippsland's long-footed potoroo (*Potorous longipes*) remained beyond human knowing. Indeed the creature was not formally described (by Victorian government scientists) until 1980. One of the first specimens discovered was lying dead on the side of a road just a few hours' drive from 20 Rose Street, Sandringham, where I lay a-turning at night, a teenager restless for adventure.

1

A Failed
Circumnavigation

Were it not for museums and their volunteer programs I probably would
have become a schoolteacher, my year-12 scorecard having shattered
my aspirations of a career in biology. La Trobe University, a newly established institution in Melbourne's northern suburbs, accepted me to
study humanities, and I enrolled in early 1974 because nothing else came
to mind. I tried to fit some basic biology in alongside studies of Greek
tragedy and the fate of the Portuguese seaborne empire, but the early
starting times of science lectures proved a fatal impediment to an
eighteen-year-old in his first year of independent living. Even with-
drawing from the course proved beyond my organisational ability, and
when I awoke on the day of the final Zoology 101 exams with the sun
high in the sky, I knew that I had failed science with the lowest mark
possible.

Specifically, it was a tall, blond Californian with a Mennonite-style
beard named Dr Thomas Rich, who had just moved to Australia to take

up a curatorship in vertebrate palaeontology at the Museum of Victoria, who sustained my slender hopes for a career in science. I first met Tom while I was in high school, and he did not care that I was a failure in the formal curriculum, but instead gave me his time so that I could learn a little, informally, about palaeontology. Early on he told me that, were I lucky enough to embark upon a career in palaeontology I had to have the will to fail, which was his way of saying that when all looks hopeless you just have to plough on. It was a valuable lesson, and without it I hate to think of where I would be today.

About this time Tom discovered that he had a problem. Some years earlier the museum had collected dozens of kangaroo skeletons preserved in cumbersome blocks of clay. Not only were they taking up valuable space in the overcrowded collection, but Tom had detected among them the plague of palaeontologists—pyrites disease. Iron pyrites form in specimens preserved in oxygen-less environments, and when such fossils come into contact with humid air the pyrites turn to sulphuric acid, causing the bones to decay to grey, puffy dust. Sometimes they even explode, creating chaos in museum dungeons. If the specimens were to be saved, the clay surrounding them had to be removed as soon as possible, and the bones coated to protect them from the air. Tom had no one to spare for the job, so he entrusted it to me.

Each Monday (a day I had no classes) I would arrive at the museum, then a grand, colonnaded Victorian edifice in the heart of Melbourne whose front entrance was watched over by a statue of Redmond Barry—the judge who ordered the execution of Ned Kelly. Tom would meet me in the echoing foyer wherein stood Phar Lap, clear me through security and usher me into the palaeontology collection. This inner sanctum was reached through creaking metal doors tall enough to admit a *Tyrannosaurus rex*, which opened onto a corridor containing an

Egyptian mummy (long superannuated from display), and a gigantic hall filled with wonders—the cast of an *Archaeopteryx*, the skull of a giant Madagascan lemur, and an ichthyosaur from Germany. Tom's office stood in a corner of this wonderland, and I would often spend an hour or so examining the eclectic specimens, many of them accumulated during Victoria's golden age in the late nineteenth century, when money for acquisitions was no obstacle. But nothing interested me as much as the fossil kangaroos, and one in particular captured my imagination. Known as *Propleopus*, it was represented in that prodigious collection by just a few teeth. But what teeth they were! Shaped like the blades of a miniature buzz-saw, it was hard to imagine how they worked in a kangaroo's mouth. What that creature ate, how it lived and indeed when it lived, all seemed to be unknown. All I knew was that long ago one had most likely made its way up the Swanston Street hill and passed the site of the museum, for several of its teeth had been found near Melbourne. The rest was a great mystery—one that, despite my poor academic record, I harboured hopes of solving.

Having exercised my imagination in the collections, I would make my way to a tiny laboratory under the stairs in the museum's dungeon, there to clean the fossil kangaroos, and thus occupied I could not have been happier, especially when Tom arrived each lunchtime to heat a can of beans and discuss fossils. Were it not for my university classes and a necessary part-time job, I would have been there seven days a week. But now the long summer vacation was looming, and I was growing increasingly indignant at the fact that even though I was native born I had never met an Aborigine nor seen the desert. So in late November 1975, at the start of the summer vacation, I temporarily set aside work at the museum, and set out to see my country.

I was immensely proud of my beaten-up old Moto Guzzi 750 sportster. I'd heard that during World War II her V-twin engine had powered Italian light-armoured vehicles across the North African desert. In my eyes she was a fabulous-looking machine with wide, sweeping handlebars and heavy wraparound wheel guards, her low tank ornamented with a hand-painted eagle. Her sleek lines and allure, I desperately hoped, would enhance my thus far slender success with the beautiful girls who stood out like Venus on the half-shell everywhere I looked. Twelve years at an all-boys Catholic school can do that to you. And it sometimes worked. I briefly befriended a lass with all the allure of a Hawaiian princess who, when riding pillion, enjoyed controlling both bike and rider through masterful application of the joystick. The only trouble was that she loved speed and a winding road, and I soon had to choose between sex and survival. Perhaps such slender victories were enough to inspire my mate, Bill Ellis, to buy another, albeit better preserved, machine. Being tall, dark and strikingly handsome, Bill didn't need a bike as much as I did, so perhaps it was a shared desire for adventure that motivated him. Whatever the case he proved to be an ideal travel partner—fearless, of few words, and remarkably tolerant of my eccentric ways. As we set off from Melbourne with a few dollars in our pockets, intent on circumnavigating the continent at the height of summer, we had no idea that we would never finish the journey.

I had decided to use the trip to collect specimens of comparative anatomy. To this end, and blissfully unaware of the need for a permit even to touch a native animal killed on the roadside, my bike was equipped with a small strap-on esky behind the pillion seat, inside of which rattled a large and gruesome-looking defleshing knife. I had thought far enough ahead to decide that I would donate any specimens collected to the museum, but more immediate issues had evaded consideration.

We headed west for Adelaide and then Perth, and it was only when we stopped, roadside, beside my first intended specimen—a splendid male western grey as large as myself—and set to work with my knife, that it occurred to me that other travellers in the South Australian outback might find such activities unsettling. Almost as soon as the thought formed in my mind the rumble of an approaching car was heard and, suddenly embarrassed at the spectacle I presented, I walked briskly away from the prone roo, whistling into the air and trying to hide the 50-centimetre-long knife behind my back. Ever tolerant, Bill agreed that we should camp nearby so I could perform the gruesome deed under the cover of darkness. After eating Irish stew from our billy I set out, parking my bike in front of the carcass with the headlight on so that I could see what I was doing. The job was made unduly difficult because I had neglected to sharpen my sabre, and after a long, bloodied struggle it became evident that to retrieve the all-important skull I would have to use the weight of the carcass to separate the neck muscles. Wet with blood and lurching under the full weight of the dead marsupial, I was so preoccupied that I did not hear the approaching rumble until it was too late. As the car accelerated past I glimpsed the family inside, horror-struck, mouths agape, staring at the frenzied bikie who was waltzing drunkenly with a disembowelled kangaroo on a lonely country road. As they disappeared into the distance I finally detached the head, after which I impinged on Bill's good humour yet again by boiling it, to remove the flesh, in our all-purpose billy.

Although our route kept us close to the coast, the green fringe of the continent soon gave way to the muted colours of the interior. It is surprising how narrow that life-giving fringe is. Nowhere in Australia is far from the outback, and every centimetre of the country is touched at some time or other by its winds, dust and flies. The flat dry inland was an utterly unfamiliar landscape, and one for which we were ill-prepared, for the Guzzis were possibly the worst bikes to take on such a trip. Mine did not even have air-filters, instead sporting elegant bell-mouths on its carburettors. But we didn't care. We were nineteen, and we were free.

On the Nullarbor, nothing among the low blue-tinged bushes stretching to the horizon stood higher than my knees. The sun baked our skin and the mirage ate up the distance, creating a sense of going nowhere. For hour after hour there was nothing but a road and a line of power poles stretching in both directions—a scar through the blue-green of the saltbush—with no sign of life.

But life there was, for the locusts were swarming. The first we came across were tiny and struck our legs like bullets—painful even beneath leather boots. The next swarm, still wingless, was larger and could leap a little higher, but their bodies were softer. The next lot had sprouted wings, and they struck anywhere. Driving into a locust cloud at 120 kilometres per hour was like driving into a living hailstorm. Any exposed skin was soon stinging with pain, and we struggled to see the highway ahead through visors smeared with the white and yellow fluid of squashed insects.

Then there was a sign: 'Head of the Bight'. We followed the dirt track, fatigued as the heat and the still, stifling air caught up with us. We got off our bikes and walked a few metres to where the endless plain suddenly ceased, as if sliced by a sabre far sharper than my own. After days of unvarying flatness the terror of the crumbling vertical cliff at our feet was compounded by the Southern Ocean, which raged with such force at its base that I could feel the shock of the waves through my boots. Its booms made me stumble involuntarily backwards to the heat, flatness and still air of the inland.

As we rode on we discovered other living things in that seemingly desolate landscape: an emu with a stately stride, a red kangaroo lying in the shade of an insignificant bush. Close to the Western Australian border, mounds began to appear. They marked the burrows of southern hairy-nosed wombats, some of which were large enough to crawl down. I squeezed head-first into one, vainly hoping to spot a wombat, and was surprised at how cool it was inside. A chance to venture further underground soon arose. Cocklebiddy Cave is a huge cavern lying beneath the Nullarbor Plain a little to the north of the road. We parked our bikes before clambering down to a yawning pit. It was an awesome space, cool and gloomy as a cathedral, but what fascinated me most was the scattering of small bones, mostly of native mice and rats, which had become extinct on the Nullarbor only thirty or forty years earlier.

We paused just east of Kalgoorlie to admire the knotted, greasy trunks of the gimlet gums, and strode over the thin crust of dried moss and lichen, which along with the last flowers of springtime suggested that this could sometimes be a gentle land. But now it was flat and dusty, the mallee a maze of uniformity where you could easily get lost. Among the knotted trunks we saw lizards and birds, and more of that subtle beauty that is so characteristically Australian—a warty grey mallee-root, a gum tree shedding its old bark in flakes and fine strips. Then, in a small clearing, we stumbled upon an arrangement of mouldering sticks on the ground, and some sturdier branches still standing. It was the remains of an ancient gunyah, though how long the bough shelter had lain decaying there we could not tell, nor could we fathom why an Aborigine had chosen that obscure place to rest. Certainly it was of a size to allow one person only in its shade.

Over the years the vision of that gunyah has frequently returned to me, and I've imagined a solitary Aboriginal hunter returning to it with a catch of rabbit-sized marsupials, to spend the night in comfort. For someone who had never met an Aborigine, and who had spent their life amid the European grandeur of Melbourne, that gunyah came as a deep shock, for it put my society in context and made the Aboriginal occupation of Australia a palpable, recent reality. I was learning that in very recent times this land had been wrested, often violently, from its original owners. And that entire ecosystems had been destroyed by sheep, the axe and the plough.

2

Captain Cook's Kangaroo

If you ever see a fresh kangaroo carcass lying beside the road it is well worth stopping to take a closer look. There is not an ounce of fat or wasted muscle on their perfectly proportioned frames, and even in death their grace and beauty—which extends from the tips of the slender limbs to their long and curved eyelashes—is sublime. But what kind of kangaroo are you looking at? This is not an easy question to answer, for the larger kangaroos belong to a group of around twenty species that are classified into three genera or subgenera, depending on whom you ask. They are *Macropus* (the grey kangaroos, whose name means 'big foot'), *Osphranter* (a name of obscure origin, for the red kangaroo and euros) and *Notamacropus*, a name given by Lyn Dawson and me to a dozen or so stripe-faced wallabies, which means 'striped kangaroo', but which is also a joke, for we wanted to emphasise that these creatures were 'not a *Macropus*', the genus in which they were once classified.

For now we will concentrate on the larger kinds—the splendid, desert-dwelling red kangaroo, the euros of the rocky ranges, and the grey kangaroos of Australia's better-watered south and east. Does your victim of the internal combustion engine have silken fur, a thick white tail and white on the sides of its muzzle? If so it is the grandest of them all, the red kangaroo, which incidentally is not always red, but sometimes grey or red-grey. If the creature has a nose like a dog, your deceased friend is a euro. If, however, it is greyish with dark tips on its tail, hands and feet, and long, straight claws on its toes, it is a grey kangaroo. But you have only just begun your identification. In travelling from Melbourne to Perth, for example, you are likely to encounter two distinctive species of grey kangaroos, while if you travel Australia's north you may stumble upon as many as three euro species.

The eastern grey (*Macropus giganteus*) can be seen in pastures and forests from Cooktown to Tasmania, while the western grey (*Macropus fuliginosus*, which can be divided into three distinct subspecies) is found from Perth to western New South Wales. In the 1980s I was fortunate enough to meet the man who announced that eastern and western greys were different. Today John Kirsch teaches biology at the University of Wisconsin in Madison, but in the mid-1960s he was an American studying in Australia on a Fulbright scholarship. John, a 'confirmed bachelor' who lives with his dog a little way out of town, describes Madison as 'The faggot buckle in America's Bible belt', a statement others may dispute, but which seemed accurate enough to me. He kindly invited me to stay with him when I visited, and was busy straining rice through an old sock (newly washed, I hoped) when we got onto the subject of grey kangaroos. The kangaroo shooters John spoke to had long recognised two kinds of greys, as had many pastoralists; eastern greys are a clear grey colour and lack a distinctive odour, while western greys are chocolate-coloured and, if male, reek of curry. This rather unexpected odour is produced from a gland on the chest, and were western greys inhabitants of the subcontinent it may well have led to their early extinction. But to many Australian nostrils the odour is overwhelming. Roo shooters call the large males 'stinkers', and avoid them when they can, and Aborigines such as the Adnymathanha of South Australia prefer other food if it is available. (It is possible that the western grey only entered Adnymathanha country in the Flinders Ranges after European pastoralists dug stock wells.)

John's principal method of providing scientific proof involved injecting blood serum from western grey kangaroos into eastern greys and vice versa, as well as injecting serum from both into possums and rabbits. The subsequent intensity of the immune response revealed how close the relationship between the creatures is. In a process known as electrophoresis, he also examined how long it took for various blood chemicals to pass through agar gel when an electric current was applied. As I listened to John I wondered at the readiness with which Australians accept the word of scientists, especially foreign experts, yet so often mistrust their own experiences and observations.

John's results were later confirmed by experiments which revealed that the two grey kangaroos had different breeding cycles and were thus unlikely to mate in the wild. Furthermore, hybrids bred in captivity had very limited fertility—rather like the sterile mule that results from a horse–donkey cross—the ultimate test of the distinctness of a species.

In truth, confusion has surrounded grey kangaroos since the time of James Cook. The first encounter between a man of science and the marsupial was the result of a terrifying accident—a holing of the *Endeavour* when it ran onto a coral shoal. In desperate need of repairs, the stricken vessel reached the mouth of the Endeavour River in Far North Queensland on 15 June 1770. Judging from the journal of Joseph

Banks, the ship's naturalist, within a week or two everyone but he had seen the amazing kangaroo. It was a frustrating misfortune, for the descriptions of sailors were often hard to fathom; one told a breathless tale of a beast 'about as large and much like a one gallon keg, as black as the devil, and had two horns on his head. It went but slowly but I dared not touch it'. Substitute the 'horns' for long pointed ears and you have a description of a black flying fox—though it is doubtful whether Banks ever got to the heart of this nautical account.

The day after hearing this tale, Banks glimpsed a creature 'like a greyhound in size and running, but had a long tail, as long as any greyhounds'. The encounter was so fleeting it left him lamenting 'what to liken him to I could not tell'. By 7 July he had decided on an expedition to settle the question and after a night spent in 'lodgins close to the banks of the river' where 'Musquetoes…spared no pains to molest us as much as was in their power', Banks rose with the first rays of the sun. His diary records:

> We walked many miles over the flats and saw 4 of the animals, two of which my greyhound fairly chased, but they beat him owing to the length and thickness of the grass which prevented him from running while they at every bound leapt over the tops of it. We observed much to our surprise that rather than going on all fours this animal went only on two legs, making vast bounds as the Jerbua does.

Saturday, 14 July 1770, was a red-letter day for both Banks and the marvellous marsupial, with the naturalist getting his first close look at the creature, as well inducting it into the English language through his journal, on a page titled 'kill kanguru'. The animal had been shot by Second Lieutenant John Gore, the *Endeavour*'s most accomplished hunter, who nearly two weeks later bagged a second specimen, this one weighing eighty-four pounds (38kg). Perhaps the kangaroo's low regard in Australian cooking originated with this superannuated individual, which Banks found to 'eat but ill'. 'He was I suppose too old' the naturalist reasoned, before recording: 'His fault, however, was an uncommon one, the total want of flavour, for he was certainly the most insipid meat I eat.'

The *Endeavour* carried three specimens back to England—including an atrociously stuffed skin—and from these animal painter George Stubbs produced the engraving that introduced the kangaroo to the world. It shows an almost cartoon-cute animal looking inquiringly over its shoulder with a catch-me-if-you-can look in its eye, and bears the following caption:

> Inhabits the western side of New Holland…It lurks among the grass: feeds on vegetables: goes entirely on its hind legs, making use of the forefeet only for digging, or bringing its food to its mouth. The dung is like that of a deer. It is very timid: at the sight of men flies from them by amazing leaps, springing over bushes seven or eight feet high; and going progressively from rock to rock. It carries its tail at quite right angles with its body when it is in motion; and when it alights often looks back: it is much too swift for grey-hounds: is very good eating. It is called by the natives, *Kanguru.*

This description is erroneous in almost every regard, confusing as it does east with west, the kangaroo's food, its use of its limbs and, according to Banks at least, its culinary qualities. The botanically trained Banks made no progress in classifying the creature and, befuddled by Stubbs' drawing and misleading caption, nor at first did anyone else. In an imaginative leap of his own, Europe's leading zoologist, le Comte de

George Stubbs' kangaroo illustration was based on a badly stuffed skin collected by Joseph Banks at the Endeavour River in 1770.

Buffon, postulated that it must be a gigantic rat, close to the jerboa or jumping rat of Africa. A Swiss naturalist then bestowed the name *Jerboa gigantea* (giant leaping rat). By the 1790s, however, these ratty affinities were beginning to be doubted by the eminent British naturalist George Shaw. Having access to better material, Shaw replaced the murine generic name *Jerboa* with *Macropus*, thus creating the binomen *Macropus giganteus* which remains the scientific name of the eastern grey kangaroo. Yet it took until 1816 for the marsupial affinities of the kangaroo to be definitively established.

Despite the immense excitement news of the creature aroused in the coffee shops of London, the physical evidence of its existence was neglected. No one, it seems, was much interested in any of the zoological specimens brought back in the *Endeavour*, and much of that priceless collection has been lost. Banks gave some of the material away (one kangaroo skull ended up with First Fleet surgeon John White), while the rest was sold at auction in 1806. A drawing of a kangaroo skull by *Endeavour* artist Sydney Parkinson has, however, survived. When it was examined by a biologist in the 1960s it caused quite a stir, for it clearly depicted the skull of a euro (*Macropus robustus*) rather than an eastern grey. The identification accords well with the otherwise baffling engraving by Stubbs, which depicts a creature with the short, curved claws of the euro on its toes, and with his description of a creature hopping 'from rock to rock' with its tail held at 90 degrees to its body, a behaviour characteristic of the euro and no other kangaroo.

The discovery had the potential to cause chaos in the scientific fraternity. Imagine the confusion if you were forced to swap the names of two favourite great-uncles—just remembering to call Cyril, Ernest and Ernest, Cyril—much less getting them to remember who was who— would have both you and them in despair. Changes in nomenclature are dreaded by everyone, but in the case of the grey kangaroo the day was saved through an unlikely agency.

The Museum of the Royal College of Surgeons is on the college's first floor in Lincoln's Inn Fields, London, and is packed with an astonishing array of specimens. Among the most popular exhibits are skeletons of the famous (including James Byrne aka the 'Irish Giant' who begged to be buried rather than anatomised), assorted midgets and the Elephant Man, all of whom jostle for space with lesser wonders, such as the skeletons of two Réunion Island solitaires (large white birds extinct since the seventeenth century), the pickled penis of a pox-struck sailor who repaired holes in his organ with whalebone, and the jaws of a wombat. But these are merely the relics of a far more wondrous compilation, for the collection was devastated by German bombs during World War II. Among the lost treasures were the twisted lower spine of Gideon Mantell, the discoverer of the first named dinosaurs (he suffered a terrible carriage accident), and the skull of a kangaroo brought back by the *Endeavour*, which was presumably purchased at auction in 1806 when Banks offered nearly 8000 lots of zoological specimens for sale. Its true identity would have remained a mystery had not someone photographed it. The surviving black-and-white prints reveal clearly that the skull was that of an eastern grey kangaroo, and this tenuous link was all that saved the species from a scientific name change.

If eighteenth-century scientists were confused as to the nature of the kangaroo, then the general public was positively baffled. The situation was not assisted by the antics of Dr Samuel Johnson, the compiler of the famous Dictionary. Soon after hearing of Banks' discovery he took a tour of Scotland, and one dreary night in Inverness enlivened the conversation with news of the antipodean novelty. James Boswell recorded how, to the astonishment of the Caledonians, the near-blind doctor gathered his ponderous, scrofula-stricken body into an erect pose, 'put out his hands like feelers, and, gathering up the tails of his huge brown coat so as to resemble the pouch of the animal, made two or three vigorous bounds across the room'.

Confusion followed news of the wondrous kangaroo even to the shores of Botany Bay. Watkin Tench, a captain lieutenant on the First Fleet, describes what happened in 1788 when the 'kangaroo' first came to Sydney Cove:

> Soon after our arrival at Port Jackson I was walking out near a place where I observed a party of Indians [Aborigines] busily employed in looking at some sheep in an enclosure, and repeatedly crying out 'kangaroo', 'kangaroo'! As this seemed to afford them pleasure, I was willing to increase it by pointing out the horses and cows, which were at no great distance.

'Kangaroo' is a corruption of a word from the Guugu Yimidhirr language of the Cooktown area and is of uncertain origin; indeed, the confusion inherent in the act of an eighteenth-century European pointing into the distance, in the general direction of a fleeing marsupial, while making culturally specific inquiring gestures at an Aboriginal man, are considerable. At various times the word has been reputed to mean 'I don't know', or 'Bugger off'. It may even be the Guugu Yimidhirr name for the euro, further confounding (if that were possible) the association of the name *Macropus giganteus* with the eastern grey kangaroo. In any case, the word 'kangaroo' was Greek to the Eora people of Sydney Cove, whose language is as different from Guugu Yimidhirr as English is from Hindi. They quite sensibly assumed it was a whitefella word for any large animal except the dog—which they taught the whites was 'dingo'.

Clarity, however, was coming. By 1791 Londoners could, for a mere shilling, satisfy themselves as to the nature of the beast by ogling a living specimen at a trunk-maker's shop (No. 31) in the Haymarket. The creature had presumably been brought back aboard some returning First Fleet vessel, and was the first of many to make the migration. By the 1820s English kangaroos were so common that you could not have got a halfpenny for a peek since herds of the creatures, along with prolific flocks of emus, had been established at Windsor Great Park and other English estates.

Today, grey kangaroos have entirely lost their mystique. It probably doesn't help that their name suggests the dullness of men in grey suits—and indeed the species remains common around Parliament House, Canberra—but as with some besuited humans, their dull exterior hides an impressive anatomy. An adult male eastern grey can outrun a greyhound or a horse, swim a mile and still have the energy to drown a harassing hound with its great hind-feet. They are superbly adapted to the well-watered eastern part of Australia, and like all marsupials are economical in their food intake. They are slow breeders, taking over eighteen months to wean their young (it takes a year for the red kangaroo), and are usually seasonal in their reproduction. But in the delicate matter of selecting the sex of their offspring the grey kangaroo shows astonishing discretion, for there is some evidence that females selectively rear daughters while they are young, and sons when they are older. The advantage the mother derives is a social one, for the bond between mothers and daughters is long-lasting and may benefit both. Sons, in contrast, wander to establish their own territory. Just how this trick is performed, however, is not yet understood.

Although eastern grey kangaroos moved westwards as stock watering points opened up, they are hopeless at coping with drought. Breeding ceases far earlier in a dry spell than with either euros or reds, and hundreds of eastern greys have been known to perish around a drying waterhole when good water and feed existed only kilometres away— something unthinkable for the red kangaroo.

Yet the eastern grey kangaroo is clearly doing something right, for as I learned when I prepared those skeletons at the Museum of Victoria, it belongs to a venerable lineage. The bones entrusted to me had been discovered in one of the biggest quarries in the southern hemisphere— the Morwell open-cut coal mine in west Gippsland. Occasionally the machinery that scoops up the coal will encounter a patch of clay, the result of ancient fires when the coal was exposed at the surface. In the 1970s a machine operator scooping up this muck somehow spotted a bone. Investigations soon showed that complete skeletons embedded in the clay were preserved in such detail that the remains of tiny joeys could still be seen in the region of the pouch. And not just that, but around the bones impressions of skin, fur and even the 'last meal' of the creatures could be discerned, the stems as green as they day they were plucked!

Such preservation is extremely rare. Nothing like it has ever been discovered anywhere else in Australia. Further, studies indicated that the remains were at least two million years old, their astonishing preservation being due to the lack of oxygen in the mud that had accumulated in ponds in the burned-out coal seam. When I studied these fossils I found that in every detail except size (the Morwell animals would have been somewhat heavier), the fossils were identical to eastern grey kangaroos still living in Victoria. The similarity even went as far as the skin impressions—the hair follicle pattern was indistinguishable from that seen in leather made from the hide of eastern grey kangaroos, but very different from that of reds and euros. I would later discover that jawbones dating back about 4 million years did not differ from those of the living eastern grey kangaroos—evidence that eastern greys had existed long before red kangaroos, indeed, before even the mighty diprotodon and giant short-faced kangaroos of the ice age.

3

Quokkas, Euros and Stinkers

After passing South Australia's Coorong on that motorcycle journey in 1975, stinkers and their chocolate females were the only grey kangaroos I saw. On nearing Perth a more experienced kangaroo watcher might have detected a subtle change in the animal—a slight lightening of the fur accompanied by a distinct widening of the foot—a sign that the boundary between the mallee subspecies of western grey (*Macropus fuliginosus melanops*), and the great southwestern kangaroo (*Macropus fuliginosus ocydromus*) has been crossed. If I had diligently hacked a foot from every road-killed kangaroo I came across on that journey and noted its age and sex, I might have been of service to science. That is because no one has yet clearly established the distribution of the mallee grey and the southwestern grey, nor determined whether these varieties intergrade or abut as distinct communities. This is a question of vital biological significance, for if they abut they may be separate species, rather than mere subspecies.

A few hundred kilometres east of Perth a sharp change in vegetation occurs. The mallee, which has dominated the landscape ever since leaving the Nullarbor, gives way to kwongan—scruffy-looking heathland where every plant seems to belong to a different species, and it is this boundary that may account for the distribution of the subspecies of grey kangaroos.

In 1975 when we stopped to examine the kwongan, venomous olive dugites with fine black spots slithered among the flowering kangaroo paws, whose brilliant green and red flowerheads shone like beacons, while every now and then a small tree smothered in orange-yellow flowers stood out above the scrub. This was the Western Australian Christmas bush (*Nuytsia floribunda*), which flowers with surprising vitality in the heat of the Western Australian summer. Only later would I learn the reason for this—it does not fuel its nectar flow with its own water but with that stolen from neighbouring plants, whose roots it taps with spigot-like structures.

By the time we were coasting down the Darling Escarpment into Perth the clutch on Bill's bike had begun to slip ominously. Back then Guzzis had a six-stud clutch—the Achilles heel of an otherwise immortal machine. Needing a replacement, we located the only Moto Guzzi dealer in the city, a man named Stolarski. But how to carry out extensive repairs with our slender budgets? Stolarski, it turned out, was the salt of the earth. Perhaps he remembered his own youth spent on motorbikes, for he suggested that we buy the spare parts and use his workshop and tools to replace the clutch ourselves. We quickly set about unbolting the gearbox and removing the clapped-out clutch plate.

Moto Guzzis are heavy bikes, weighing over 200 kilograms, and their stands consist of a band of metal with a thin keel running along its underside. We were black with oil and grease by the time I eased the repaired bike off its stand, in the process crushing my big toe between the metal keel and the concrete floor. As Stolarski watched me hop around the workshop, blood flowing from my running shoe, he reached for a huge spanner and bashed it against his right shin. 'Crushed your toe, eh? That's nothing,' he said, as he hoisted his trousers to reveal a false leg. 'I lost this in a racing accident years ago.'

A none-too-pleased matron at casualty screeched at me not to touch the walls, insisting I wash my filthy hands if I wanted to be examined. An X-ray revealed no broken bones, so a few days' rest and the toe would repair itself. As Bill and I had no money there was only one thing for it: we left the bikes at ever-patient Stolarski's and headed for Rottnest Island to sleep rough until the flesh healed. Perhaps we would find oysters and fish.

Rottnest is a low, limestone island lying a few kilometres off the mouth of the Swan River and has long been a holiday playground for the citizens of Perth. Bill's parents had honeymooned there and apart from an unfortunate incident involving Bill's dad snoring open-mouthed on the beach and a defecating seagull with a crackshot aim, they had come away with wonderful memories. But I was at best a hobbler, and we soon discovered that non-paying campers were expressly forbidden by law. It was a bleak place to try to hide, for there were few thickets and the best I could do was to hop to a clump of melaleuca bushes near the beach. There we set up camp and watched the army training a hundred metres away.

That night I was awoken by a sharp thump by my head. Fearing that our hideout had been discovered I peered up cautiously to see two beady eyes observing me at close range. It was a quokka (*Setonix brachyurus*), a member of the kangaroo family for which the island is famous. These cat-sized creatures had constructed a quokka highway under our bush, and my body lay right across it. Throughout the night these compact, brown wallabies, with their trusting faces and short tails, thumped the ground in alarm as they confronted the unexpected obstacle.

At the time I saw quokkas in 1975, the species was at the nadir of its existence. They had all but vanished from the mainland, their last secure strongholds being Garden Island and nearby Rottnest. They were one of the first Australian species to be described by Europeans. The Dutch navigator Samuel Volckertzoon landed on an island in 1658 and described wild cats, with 'pouches below their throats into which one could put one's hand, without being able to understand to what end nature had created the animal like this'. In 1696 another Dutch navigator believed the animals were a kind of large rat, and named the place Rats-nest Island (now Rottnest). Another claim to fame is more recent, dating to the 1950s when investigation of the quokka's reproductive system led to the discovery of a phenomenon known as embryonic quiescence. This remarkable strategy, whereby mothers can suspend the growth of their embryos for up to a year, has proved to be widespread among kangaroos, and is one of the key factors in their success; but more of that later.

After a few days of saltwater bathing had done its best for my mangled toe, Bill and I returned to our bikes, intent on resuming our circumnavigation. We headed north, up the west coast of the continent. The landscape beyond Shark Bay was hallucinogenic—blood-red, yellow and white mesas that, in the heat-haze of summer, hovered above the horizon. With the stifling heat and monotone rhythm of the V-twin engines other sounds became muted, and we were soon gliding along in an ethereal world of shimmering, illusory shapes and colours, like satellites through space.

We drove past the Murchison District in Western Australia's mid-north without giving the place a moment's thought. It would be almost a quarter of a century before I would have another chance to explore it.

In 1996 by way of a letter from Sandy McTaggart, secretary of the Murchison District Sports and Shooters Association, asking if I would address their annual general meeting. When I phoned Sandy to find out more he mentioned the other organisers, Jock McSporran and Muggon Bill. The names had me wondering if this was an elaborate practical joke, but then Sandy told me that Mount Narryer lay on his property, and I was hooked. Mount Narryer is home to some of the oldest known rocks on our planet—4.2-billion-year-old zircons.

The Murchison sportsmen and shooters hold their annual meeting in spring, when for three weeks their country is transformed into a technicolour carpet of daisies and other wildflowers. As I drove up the Butcher's Track towards Sandy's homestead, drifts of paper daisies painted the roadsides pink, yellow and white. As beautiful as this was, it formed at best a threadbare carpet that could not hide the degraded condition of the land. This was marginal sheep country, and overgrazing had reduced much of it to eroded sand and clay, while foxes and goats were present in such plague proportions that even the rabbits seemed under threat of extermination. I later discovered that dingoes had been hunted out well beyond the dog fence, which in this area was abandoned in 1952. The dog fence is a barrier designed to stop dingoes moving into the agricultural and sheep districts. Cattle can graze outside the fence, for cows are generally too large to be worried by dingoes. Sheep and dingoes, however, do not mix, for a dingo can kill hundreds of sheep in short order, destroying a grazier's livelihood in a single night. But here, in the absence of Australia's top predator, mangy, half-starved foxes and goats had increased almost unchecked.

Dressed in a broad, battered hat, boots, and a belt sporting a prominent buckle, Sandy personified the Australian grazier. His arms and hands looked strong enough to hold down a bull, but like so many who drive everywhere in this extensive country he was a little overweight and underdeveloped in the leg department. It may have been just the type, but I felt sure that I had seen him before. A cold beer was thrust in my hand as I entered the kitchen and met the family, and Sandy was soon explaining how he liked to drink champagne—the $3.50 per bottle Great Western variety—and how I'd always be welcome if I brought a case, when I saw the *National Geographic* magazine in pride of place on the table. There, on the cover of an issue that celebrated Australia's bicentenary, was Sandy McTaggart and son—symbols of the nation.

People cling proudly to their traditions in the Murchison, but it was clear that the end must be near for much of this clapped-out land, at least under its present use. Many graziers had been reduced to harvesting feral goats to pay school fees and put bread on the table. And such were prices then that a goat destined for the Middle East live meat trade fetched more at the dockside than a fat Murchison wether. I was particularly interested in Sandy's views on kangaroos, as I felt that a sustainable harvest of them might ease grazing pressure on the overcropped pasture and add another line of income to the local economy. To my dismay, however, Sandy told me that the red kangaroo—the mainstay of the shooting industry in Australia's drier regions—was now uncommon in the district, while its smaller and less valued cousin, the euro, remained abundant.

The euro or hill-kangaroo is the most widely distributed of all kangaroo species. Except for the cold southeast, it is found wherever suitable habitat exists. If you have ever seen a male of the eastern subspecies (also known as the wallaroo), which inhabits the Great Dividing Range, you are unlikely to forget it. They are great, black, shaggy, muscular beasts resembling the mythical yowie as much as anything else. With its massive forearms, short ears and a propensity to hop in a more upright posture than other roos, a male euro moving through its hilly habitat is doubtless the origin of at least some yowie sightings—by 'jolly' swagmen and inexperienced bushwalkers alike.

Away from the Great Dividing Range the euro varies greatly in appearance. Around Broken Hill in western New South Wales they are often reddish-black, while in Western Australia the coat is a rich mahogany, and less shaggy. The females are usually lighter in colour than the males, and only half their size. Among the euro's greatest assets are the soles of its feet, which possess a most magnificent pad, so expansive, tough and well buffered by an underlying layer of fat as to resemble a pneumatic tyre. Thus armed, it can hop effortlessly over the rockiest of hillsides.

Western Australian sheep-men have long harboured a deep antipathy for the euro. Throughout the 1950s and 60s—that golden age when wool was worth a pound per pound and merino rams rode in the front seats of Rolls Royces—the graziers of the Pilbara (a region lying to the north of the Murchison District) were missing out, for their flocks dwindled while their paddocks filled with euros. Certain that the euros were pushing the sheep out, they called for scientific help, and Tim Ealey, a student at Monash University, was sent to investigate.

Known to his colleagues as 'the eurologist', Ealey is a singular personality, well equipped to deal with months of field work in difficult country. He discovered that euros do not need to drink most of the time, and when they do they can get by on less than half the water needed for a sheep or goat. Part of their secret lies in not letting any moisture out. Euro scats are so dry (among the driest produced by any creature) that you can light a fire with their turds *de jour*—and they are miserly pissers as well, possessing the ability to recycle urea through saliva (and thence to the gut), thereby avoiding the necessity of urinating. Their best trick, though, is being able to survive on the least nutritious of vegetation, the spiky spinifex grass that is at once an ornament to, and the horror of, the inland. (You will only understand the horror of spinifex if you have tried to walk through it—suffice to say that it can turn even a camel's legs black and hairless.) So efficient is the euro's gut, and so limited the creature's energy needs, that if cardboard boxes grew on bushes the euro would likely thrive by eating those as well.

Armed with this knowledge it was easy for Tim to see why the sheep of the Pilbara had ceased breeding and were withering away from starvation, while the euros continued to multiply. But Ealey found that the euros were not replacing the sheep without assistance—the sheep-men were helping them. For decades the graziers had been severely degrading their pastures through overstocking. As a result, sheep and even red kangaroos were unable to sustain themselves and the euros, which had once kept to the sterile hills, now inherited the blighted plains. Thus the rise of the euro had not been brought about by its superior competitive abilities, but by a wholesale collapse of the ecosystem.

I explained to Sandy that an abundance of euros away from the rocky ranges is a fair indicator that the country has been well and truly flogged. I then went on to say that the scenario that had occurred in the Pilbara decades earlier was now playing out further south. We both knew that the sheep industry was dead in the Pilbara, so the conversation was not an easy one. Tragically, the situation meant that not even native marsupials can contribute to economic viability—the reds are gone and the euros are too small and elusive to be attractive to roo shooters.

My speech at the meeting the following morning did not go down well. Most of the audience lived with the problem of environmental degradation every day of their lives, and they hardly needed an outsider lecturing them on the finer points of the issue. Question time began when a grazier who resembled an overwound spring leapt to his feet and exclaimed that city experts were always giving out useless information, and the best thing would be if they all took a half cut in pay. When a second questioner accused me of expecting them all to live like savages, Muggon Bill rose to save the day. 'Time to open the bar,' he said as the clock struck 9 am. In the more convivial atmosphere that soon developed, a group of graziers spoke of the difficulty of their situation, and of their eagerness to conserve the native species that remained on their properties.

That evening we regrouped at the big tent in the showground for the annual dinner. After the main course Jock McSporran (who looked remarkably Scottish) announced that well-known local identity Luigi the shearer would entertain us with a rendition of 'Ave Maria', 'O Sole Mio' and 'Back to Sorrento'! Shouts of approval shook the tent as Luigi, a portly Italian whose tenor voice would have been passable in any shearing shed, rose to his feet. I respectfully knocked back a few beers as he sang, after which the MC informed us that young Jacko would continue the entertainment with a rendition of Tina Turner's classic 'Simply the Best'.

I was sure that I'd had a beer too many when I saw a young black man dressed in drag slink onto the stage, a mophead perched on his cranium and balloons for a bosom. He got a few laughs as he strode around, shaking his extravagant breasts and striking the air with a clenched fist. But the best was yet to come. After a quick costume change our singer re-emerged as Al Jolson. He burned through 'The Old Folks at Home' and was starting in on 'My Mammy' when he sidled up to the only well-dressed couple in the audience. Clearly the local aristocracy, the man looked very pukka in his tie and jacket, while madame's blue rinse and evening dress would have been respectable in Perth Casino. They were the only people not laughing as 'Al' approached. Indeed the woman looked close to panic as Jacko got down on one knee, looked lovingly into her eyes and crooned; 'Mammy, Mammy, I'd walk a million miles for one of your smiles, My Maaaa-mmy!' It was a finale that brought the house down.

Sandy had promised to take me out to Mount Narryer the following morning, and as we drove we came across the most magnificent brahmin bulls I'd ever seen, their great grey bodies shooting out of the scrub in a cloud of red dust as they wheeled to challenge us. 'That's Dishwasher,' said Sandy's wife Carol, pointing, 'that's Hoover, and that's Holiday.' The family had decided exactly what they'd bring when they were sold. 'Jeez, they've grown up quick,' said Sandy in quiet awe. 'Well overdue for the doctor,' he added a little nervously. As I watched a tonne of muscle pawing the ground in defiance, I wondered about the wisdom of approaching those creatures with the intent of robbing them of their masculinity. It was also a lesson for me on how hard it sometimes is for someone like Sandy, who is a sheep-man to his boots, rather than a cattle-man, to diversify his business.

For me the journey to Mount Narryer was a spiritual excursion. I had my children with me (Emma aged ten and David aged twelve) and wanted to show them this special place. Sandy explained that he had bought the pastoral lease over the Mount Narryer block as an anniversary present for his wife. She had an interest in geology, and land values were such that it had cost him less than a decent diamond ring.

Soon we emerged into the stony country surrounding the low range, and before us an undistinguished hill rose above the mulga. Red-brown and eroded into boulders on top, it had stood above the surrounding plain for eons. More than a trillion sunrises and sunsets had played over it, and it had watched over the creation of life, from humble bacteria to humanity itself. The rock that formed the hill had solidified before the atmosphere was breathable, before the sea was blue and the land green. And inside that rock were zircons dating back 3.9 billion years—close to the time Earth formed. Here, confronting myself and my children, was the yawning chasm of time, as wide and blue as the Murchison sky and as endless as the surrounding arid plain. I have never felt so insignificant, nor so correctly calibrated against the universe.

4

The Last of the Frontier

In 1975, though, it was Port Hedland that loomed on the horizon, and with it the greatest challenge for our old roadsters—1200 kilometres of dirt road leading to Halls Creek. Although it gloried in the moniker National Highway No. 1, the road turned out to be a nightmare of bull-dust and corrugations: the Guzzis were soon shaken apart as screws dropped out, glass exploded into shards and the heavy fenders threatened to snap in half. There was just one petrol stop in the 600 kilometres between Port Hedland and Broome—the Sandfire Flats roadhouse. We'd seen it signposted and, exhausted and low on fuel, we hauled ourselves up its long dirt track. It consisted of a couple of shipping containers and a metal box structure for a bar, plonked down in the sand with a petrol bowser in front. Out back lay a graveyard—hundreds of cars, trucks and motorbikes that the road had beaten. The entire place was as dismal a purgatory as any I had seen and was also, we would learn, as difficult to escape as the real thing.

As we pulled up to the bowsers a distracted-looking man standing by his car approached us and began pleading for help. He had arrived two days earlier, he said, and needed petrol to continue south. It had been 45 degrees Celsius in the shade when he arrived; the owner, let's call him Barry, began to fill his vehicle while smoking a cigarette. When the customer had mumbled something about a fire hazard, Barry had become irate and announced that he wouldn't serve the 'bloody whinger' at all. For three long days he had proved as good as his word. We had, I concluded, reached the frontier.

We found Barry in the airconditioned interior of the metal box, propping up the bar. He was a man for whom, judging by his ravaged features, the sun was always below the yardarm; and Sandfire Flats was a terrible place to have gone to fat. Photos on the wall showed the place when all that could be seen in the sea of scrub was a truck with a 44-gallon drum on its tray-back with a younger Barry standing atop it, serving fuel with a hand pump.

As we finished our beer in the shade of the petrol bowser's pergola, a thin man with a pinched face sidled up to me. He said that he was being paid to paint the shipping containers. He'd been there for months but he hadn't done very much because he was paid in beer. We told him that we were riding around Australia. 'Watch out for those bloody donkeys,' he said. 'They stand with their arse to yer, and yer can't see them because their stripe looks like the line down the middle of the road. Hit one and you'll have a face full of donkey-arse at one hundred kilometres per hour. And watch out for the bloody Abos too. Can't trust the bastards, mate. They'll do you if they can!'

That evening Bill and I rode on into the waning light for an hour or so before pulling off into the mulga, still over 200 kilometres from Broome. We travelled light so camping was a pretty rough business. On this night we stretched out our pieces of foam and lay under the stars, me sleeping in an old World War II dispatch rider's uniform I'd picked up at an army disposals, Bill in a light sleeping bag. We dropped off to sleep straight away, but were awoken around midnight by voices close at hand. People were talking excitedly in the harsh, guttural tones so characteristic of the Aboriginal languages. My imagination ran riot, and I was instantly gripped with a fear that Europeans have experienced ever since the first explorers pushed into the tribal lands of Australia. Had they found us, and if so would they attack us?

A few minutes later it became clear that they had no idea that Bill and I were just a few yards away, and my fear began to abate. It must have been an uncanny coincidence that they had pulled off the road just where we had. We figured that it was only a matter of time before they discovered us, so I decided to take the initiative. With my heart pounding, I placed our largest shifting spanner in my back pocket, and walked towards the voices.

Two old Holden station wagons were pulled up by the light of a newly kindled fire, and men, women and children huddled around a billy. An old man, who was talking to a youth holding a tyre lever, cut his conversation mid-sentence as he saw me approach. I was trembling with fright at this, my first meeting with Aborigines, and their sudden silence and shocked looks did nothing to reassure me. The old man pointed to the flat tyre and mumbled that they had been to a wedding at Meekatharra and were on their way back to Broome. 'Oh…right. Good night,' I replied, while trying to act as if it was the most normal thing in the world for us to meet that way, at such a time and place. As soon as I walked into the darkness they took off without fixing their puncture or enjoying their tea. It was a long time before I could sleep, kept awake with fright and thinking about those people we had so unexpectedly met. And why the almost primal fear?

Just before dawn I was awoken yet again—this time by pain in my right shoulder. I thought it might be a snakebite, then discovered that the culprit was either a scorpion or centipede, for the wet season was on its way and the ground was covered with their tracks. The pain got worse, and with it came a paralysis of my right arm. It was clear that I could not ride that day, so we hid my bike in the scrub and I rode pillion on Bill's. Eighty kilometres from Broome Bill's bike began to splutter. We had run out of fuel. We waited a couple of hours beside the track, a few cars and trucks whizzing by before an old tray-back hove into view and stopped beside us. It was filled with Aboriginal people—mostly young men—smiling and laughing in anticipation of a visit to town. They told us that they had come from La Grange mission, and they not only took us aboard but hoisted our bike onto the tray as well.

On the outskirts of Broome our truck passed one of those sights that, in the heat of the desert, you're not sure you've seen at all. An old man, whose intensely black face was framed in a mane of pure-white hair and a great, bushy white beard, and dressed in what looked to be a white dress-shirt and black trousers—the remnants of a tuxedo perhaps—was riding a bicycle. As we roared past he became engulfed in a cloud of dust, and looking back we saw that he had ridden into the gutter where he stood shaking his fist at us. His name was Lawrence Williams, a young man in the tray-back told me.

By the time we got to Broome I felt sufficiently recovered to skip the doctor's, and we wasted no time completing our business and making our way back south. But when we arrived at our old campsite we were aghast to discover that my precious Moto Guzzi was gone. Stolen! Who would have done such a thing? Despairing, we continued on to Sand-fire Flats to ask Barry if he knew anything. 'It's the bloody Abos, mate,' he said flatly after I explained what had happened. 'They'd steal any-fuckin-thing.'

I sat in deep shock, uncertain of what to do, when the drunk who had warned us of donkeys sidled up to me. 'I'm getting out of here, mate,' he confided, then walked towards a wreck of a car—*sans* doors and bonnet—a few metres away. It evidently also lacked a starter-motor, for beside it stood two withered Aboriginal women whose job it was to push-start the vehicle. Before the machine jumped to life with a ghastly clatter and shot off in a cloud of black smoke, the drifter turned to me and said, 'Have a look round the back.'

I wasn't sure what he meant, but I started to wander in the square-kilometre graveyard of wrecked vehicles, and before long saw the distinctive blue shape of my bike's fuel tank. With all the outrage a young man can muster I confronted Barry with this evidence of his perfidy. 'Oh, you mean *that* bike,' he said with a smirk. 'I'll have to charge you towage for that. If I hadn't brought it in the Abos would have got it, for sure.' Feeling angry and foolish beyond reckoning, Bill and I bolted without paying the mooted towage. A kilometre or so out we saw the drifter, his car dead by the side of the road, walking slowly back towards Sandfire.

Every Australian outback town seems to have a caravan park, and such is their allure, or perhaps economy and convenience, that a wide cross-section of humanity can often be found there. Kombi-driving hippies, grey nomads, washed-up workers doing it tougher than you would think possible, they all converge in these homes away from home. Perhaps caravan parks are the last refuge of the Australian ideal of mateship, where high and low, fortunate and forlorn, gather together in a common quest for shelter and where, on 25 December, a motley assortment of Austral humanity gives thanks for the birth of our Saviour. The beachside Broome caravan park, filled with mango trees dangling delicious fruit, looked like a paradise when we arrived a few days before Christmas 1975.

The incessant humidity and afternoon rumble of thunder over the ocean proclaimed that the wet season was imminent, so we decided to travel on to Darwin in convoy with others who would leave after Boxing Day. We were, as the old bush saying goes, 'close up flyblown'—flat broke—but a phone call to Mum saved the day and our budget once again stretched to the odd beer. So we went to one of Broome's two pubs to celebrate reaching the halfway mark of our circumnavigation. We had been on the road for a month.

Broome had not been discovered by tourists then, and was still a tiny settlement with a neglected air. A few pearl luggers lay in the mud at low tide by the wharf and the beaches were mostly deserted—on some there was no one for miles. At the caravan park we learned that Broome was a magnet for people fleeing either the long arm of the law or angry wives—the kind of place where nicknames abounded while surnames were hard to learn. The town's few shops had the feel of the Chinese trade stores that are scattered through the Pacific—piles of goods here and there, tins of steamed duck resting alongside rubber thongs on rudimentary shelves.

The pubs seemed to be colour-coded—one for black and one for white—though entry to the white pub seemed more tightly controlled. Standing on the balcony of the black pub we spied the ebony face and white mane of Lawrence Williams. 'Can I buy you a beer?' was all I could think of by way of introducing myself. Lawrence accepted and we started to talk. I was captivated by his voice, for he spoke the distinctive 'pidgin' of Aboriginal Australia, but with an unmistakable English plum in his mouth, making his voice as incongruous as his appearance.

Lawrence had been born and educated on a mission north of Broome where a Church of England pastor and his wife spoke the Queen's English—which explained why Lawrence's voice called to mind the BBC World Service. Among his most vivid memories was the bombing of Broome on 3 March 1942, when the Japanese destroyed a fleet of flying boats that had evacuated Dutch women and children from Java. Lawrence had walked the beach for days retrieving mangled bodies. In his youth he had also hunted dugong. 'You know, dugong just like a lady,' he said winking at me. I didn't know what he meant, but learned afterwards that the genitals of these large marine mammals are as similar to those of a female human as nature ever invented. As the days passed we often met Lawrence around town and stopped to chat. I soon came to appreciate that the first Aborigine I ever became acquainted with was a fine person as well as an endless source of knowledge.

Christmas Day dawned to an unaccustomed bustle of activity at the caravan park. A sort of ladies auxiliary had been formed which had, the day before, been busy making devilled prunes, sausage rolls, salads and roast chickens. Some generous soul (perhaps a fisherman) had donated a huge basin full of cooked prawns, and various bottles of celebratory lubricants had found their way into the open. The men were at work erecting and laying out the tables that had been gathered from every corner, and upon which bowls, plates and bottles were placed. The early start was needed to create the right atmosphere for the great outdoor celebration and, in deference to the weather, it would be more of a brunch than the typical southern Australian afternoon affair.

After surviving for some days on mangoes plucked from the park's trees, the Christmas feast was looming large for Bill and me. As the morning ripened to a sticky humidity we stood ready to contribute our carefully conserved beer and packet of nuts to the communal feast. Then an early reveller, clad only in a dubious-looking pair of shorts, stumbled out of a dilapidated caravan. He was holding a lemonade bottle filled with an orange liquid. 'Happy Christmas. I'm Paddy,' he said in a thick burr as he stumbled towards us, already the worse for wear.

Paddy had the solid jowls and sagging lower eyelids of a bulldog, and by the look of him had led a 'terrible hard life'. He told us that the Melbourne Waterside Workers' Federation had sent him to Broome years earlier with a brief to unionise the waterfront. 'I'm the only communist in Broome,' he lamented, 'and every bastard is trying to kill me. See the doctor over there?' he said, pointing to a middle-aged, neatly dressed man approaching the feast tables. 'I'm the luckiest man alive escaping that bugger. Last time I was in hospital he tried to kill me. Next time he'll do for me for sure.'

Paddy's problems, it transpired, stemmed from his discovery that nearly everyone on the Broome wharves was an illegal immigrant. They ran a mile when the union was mentioned, leaving Paddy with nothing to do but replace his precious bodily fluids from a bottle of booze. The caravan park had been his home for over a decade.

As the gathering grew Bill and I gorged on devilled prunes, chips, sausages and salad, and we started to meet more of the locals. One Aboriginal woman wandered through the crowd, vacant-eyed, nursing a massive, raw wound on her hand—as if her thumb had almost been severed. Shocked at her injury, I tried to find the man Paddy had identified as a doctor. All he said in response was, 'Old Pol. Alcoholic prostitute. We see her almost every week at the hospital. No way I'm treating her today—she can bloody well wait until after Boxing Day.' I was shocked by his callousness, but worse was to come.

By afternoon only a hardened core of caravan-park locals remained, and I found myself talking to a wizened, grey-haired bloke with a wild look in his eyes. He claimed he had once been the local jailer, but had just emerged from a stint inside himself. 'The fuckin' magistrate put me away for three months,' he whinged. 'Reckoned I failed in me duty. All I did was let the fellas out on a Saturday night. They all come back by Sunday morning.'

'Did you let everyone out of jail?' I asked incredulously.

'What d'ya think I am! A bloody idiot?' he screamed. 'I didn't let the bloody Abos out!'

As lightning flashed on a darkening western horizon, Paddy once again sought out our company. 'Come back to me caravan,' he said, 'and have a little tipple.' The place was an appalling mess, foul-smelling as only an unclean tropical habitation can be. Paddy opened his fridge and cockroaches cascaded out from around the door. Inside was nothing but two wine flagons filled with the same sickly orange liquid. 'What is it?' I asked.

'It's me rocket fuel—methylated mango-pulp,' he said proudly as he picked up three grubby glasses.

I did my best to be social while not drinking the stuff. Paddy fumbled for a packet of old photos, sorted through the faded images and began a well-rehearsed monologue. 'This is the wife that left me,' he moaned and passed us a black-and-white print of a portly woman in 1950s suburban attire. 'And these are the children I've never seen for years,' he added and passed us another of two toddlers. He was building to a full-blooded keening as he came to the last photo, which showed a small terrier sitting outside the caravan. 'And this is the dog that left me,' he wailed, before shuffling through the entire stack again. Bill and I made our excuses and left, though Paddy hardly seemed to notice.

That evening we found Lawrence Williams at the blackfellas' pub, sitting on the steps, surrounded by drunken revellers. 'Happy Christmas, Lawrence,' I said. To my horror he rounded on me, fist drawn back. 'You fuckin' bastard,' he roared. 'I'll *kill* you.' Still not realising the danger I was in, I said 'Hey, it's me!' and pushed my can of beer into his hand before backing away. He looked confused, but still murderously angry about something. And in my innocence I had no idea what it was.

I understand better now. What I didn't know in 1975 was that for many years people like Lawrence Williams had been denied both respect and fundamental human rights. I don't know what conditions Lawrence endured on the mission, but I do know that many such places were so paternalistic in their approach that they robbed proud men of their dignity. Until the 1960s many Aborigines had no right to vote, drink in a pub, work under union awards or travel off their reserves without a pass, and until 1967 they had not even been counted in the Australian census. In short, they were considered fit to fight for Australia in the armed services, but not good enough to count as Australians.

Looking back it seems strange that I remember vividly the Cuban missile crisis and the assassination of President Kennedy, yet I have no memory of that 1967 landmark referendum. It was if a cloak of embarrassed silence had enveloped my white society—the outcome of a deep shame rooted in the unspoken knowledge of our treatment of 'them'. Nor did I realise how close those frontier days of violence between black and white—when guns and spears spoke a common language—were. Within a few hundred kilometres of where Lawrence and I stood, Aborigines who had never seen a white man still held their country. I would be twenty-two when the last of those traditional Aboriginal nomads walked in from the desert.

5

Of Nailtails and Nailed Tyres

The wet season had set in by the time we headed for Halls Creek. On the first afternoon the sky darkened as if bruised and battered, then a tropical storm burst upon us. We drove on for five minutes or so, but when the rain became so torrential that I could barely see my handlebars, our convoy stopped and Bill and I took shelter in one of the caravans. It was all over in an hour or so, and the thunder and lightning that had hushed all conversation drifted with the clouds into the distance. As the rain cleared I could see the worn landscape of the southern Kimberley, the rocks resembling ruined castles with water rushing out of every crevice. Our vehicles were in the midst of what now resembled an inland sea. Nearby, only a line of great, silver baobab trees—still to sprout their leaves and doubtless marking a creek line—could be seen above the water. It was time to stop for the night.

The following morning progress was excruciatingly slow. The inland sea had receded, but every floodway and creek crossing was running and these low points were floored with concrete that had sprouted a coating of slime—deadly for motorbikes. When the dirt road passed through a patch of scrub a lesser hazard presented itself in the form of a lithe, muscular shape that darted out onto the road almost under my wheels. It was a small, sandy-coloured wallaby, and it set my heart on fire, for a peculiarity of its gait revealed it as a great rarity. The grunter, also known as the organ-grinder or northern nailtail wallaby (*Onychogalea unguifera*), has a characteristic hopping posture, its left paw placed somewhat in front of the right, with elbows out as a bouncer might if holding two men in headlocks. The creature seems to work hard as it gets along, for its arms rotate and it emits an audible grunt with every leap. Just why it behaves in this way is not known, in fact very little is known about the creature's ecology and evolution. Aboriginal hunters knew it well, though, indicating its lopsided gait (which extends to one foot being placed slightly in advance of the other) in their sign language. If a man, stalking through scrub, held his two index fingers out, one somewhat in front of the other, you knew him to be on the lookout for a small, solitary kangaroo which can explode from under a bush with the velocity of a rocket.

In 1981, in the Northern Territory, I would make the acquaintance of this creature again, but this second meeting had less favourable consequences for the wallaby. A grunter dashed out from the scrub and before I could swerve it fell under the wheels of my truck. Although dismayed at the collision, it provided my first opportunity to examine it in detail. Except for its long, whip-like tail, it looked like a small agile wallaby (*Notamacropus agilis*). The tail has a black crest of hairs on its upper surface, culminating in a black tuft, in which is embedded a flat, black nail the size and thickness of my thumbnail. Just why nailtail wallabies grow such a remarkable excrescence at the end of their tail is still unknown, as no one has ever seen it put to use.

I scooped the body onto the back of the truck, and back at camp I dissected its unusual feet to see what could be learned of their anatomy and function. They are built for extreme speed in a forward direction and, like a horse with its single hoof per leg, the side toes were almost vestigial. Curious as to the culinary potential of the species, I took a few steaks from the haunches and made a stew. It was delicious—far superior to red kangaroo—with a taste resembling steak and mushrooms. To my annoyance I discovered on emerging from my tent the next morning that birds of prey had stolen the carcass, depriving the museum of a specimen and me of further investigation.

While the northern nailtail can still be seen in Australia's tropics, the two other nailtail species have not fared as well. The hare-sized crescent nailtail wallaby (*Onychogalea lunata*) once abounded in the west and centre of the continent. It had the unique trick of escaping predators by climbing the inside of a hollow tree, much like a sweep going up a chimney, and emerging at a spout high above to keep an eye on its pursuer. This was an insufficient defence, however, against habitat degradation, and by the 1930s the species had all but vanished. The last one ever seen by a European was captured around 1927 by Mr W. A. Wills, a 'dogger' (dingo trapper) working on the northern Nullarbor Plain, who sent it to the Australian Museum in Sydney. In the early 1980s scientists Norm McKenzie and Tony Robinson undertook a survey of the Nullarbor Plain's fauna. They were delighted to learn that Wills was well known in the region and still alive. Excited at the prospect of learning something about the extinct wallaby from one who had seen it in life, they tracked him down to an aged care home in Albany, Western Australia and swiftly made an appointment to visit the nonagenarian. Upon arriving, however, they were dismayed to learn that Mr Wills had departed the evening before. Evidently nervous at the prospect of a visit from government officials, the old fellow had fled into the night in his Holden ute, driving right across the outback to seek refuge with his brother in Queensland!

The story of the bridled nailtail wallaby (*Onychogalea fraenata*) of inland eastern Australia is only a little less tragic. This animal is larger than its western relative, around the size of a collie dog, and with its boldly striped shoulders, hips and face, it is one of the most beautiful members of the kangaroo family. It once abounded throughout the inland agricultural zone from Queensland to Victoria, where it was known as 'Flash Jack', perhaps because of its speed and striking colours. Its pads, crisscrossing the scrub and plains in all directions, possibly gave Australians the expression of 'taking to the wallaby track' or 'taking the wallaby' as slang for travelling the remote outback has it. In 1862 the biologist Gerard Krefft described it as 'the most common of the smaller species of the kangaroo tribe', yet by the 1930s it too was widely believed to be extinct. Very few vanished species are given a second chance, but the bridled nailtail wallaby is one, and the circumstances of its rediscovery are as unusual as they are fortuitous.

The story began in 1973 when a fencing contractor, Mr Challacombe of Duaringa, was called to fence an area of brigalow scrub near Dingo in central Queensland. In those days pastoralists were given incentives by the Queensland government for clearing scrub, and could even lose their land if they did not 'improve' their holdings by clearing a percentage of it on a regular basis. This property was one of the last in the region to go under the ball and chain—the cocky who owned it perhaps reasoning that the yield from its poor soil was not worth the effort. Or perhaps it was a case of *mañana*. Whatever the reason the fatal hour was now at hand, and fencing and clearing had to be done come what may. Mr Challacombe was—strangely perhaps for one of his profession—a devoted reader of *Woman's Day* magazine, its pages filled with handy recipes, knitting patterns and hints on maintaining a happy marriage. The issue Challacombe carried that day happened to include an article on Australia's extinct animals, and he noticed a striking resemblance between an 1840s painting of the bridled nailtail wallaby in the magazine and the creatures hopping through the soon-to-be-cleared brigalow.

Challacombe spoke to a National Parks ranger, and a professional investigation confirmed his identification. It was clear that the property had to be purchased to preserve the creatures, but that was not an easy process, for the discovery had instantly converted the most backward property in the region to a unique conservation asset of the highest value. Eventually a price was settled on which, it is rumoured, was sufficient to allow its owner to retire to the Gold Coast, there to lead the life of Riley. The wallaby too is now prospering, and plans are afoot to reintroduce it to parts of its former range.

After my first encounter with the northern nailtail wallaby, the journey towards Darwin just got worse and worse. The road was in awful condition, with huge potholes everywhere. We arrived in Fitzroy Crossing, still around 1000 kilometres from Darwin, covered with grazes and wet to the bone. The 'Crossing' I encountered that evening is something I wish never to see again. We stopped at a pub which was surrounded with makeshift camps. Some Aborigines were living under traditional gunyahs, while others had only canvas, or even the upturned bodies of burned-out cars over their heads. As I walked towards the post office an old man scrambled out from under a canvas sheet and asked, 'You want woman? Ten dollar—hey, you! Just ten dollar! I got two women.'

That evening I lay under the stars in my dispatch rider's uniform, listening to the shouting, screams and thumps in the drunken camp, but a few hours later the chaotic sounds began to be replaced by a low, rhythmic chanting. Then the throbbing of the didjeridu rang out, accompanied by the clicking of full and empty beer cans (replacing the clap-sticks of earlier times). I could see young men, drunk but dancing, moving in and out of the light of a big fire. The Cobba-Cobba, as cor-roborees are known in the northwest, went on until the small hours.

Even today it is hard to convey the mix of emotions that Fitzroy Crossing brought out in me: despair at the utter degradation of the Abo-riginal people gathered there; amazement and admiration at how they clung to their culture; hatred of the publican and my own society that was living off their misery. And, most of all, confusion. How had this come to be?

Before I could grapple with this question our convoy left the place in the brightness of a wet-season dawn. We had not gone fifty kilo-metres when my bike suddenly yawed on the road and skidded to a halt, its rear tyre shredded by sharp iron lying on the track. There was no choice but once again to hide my beautiful machine in the bushes, and continue on the back of Bill's to Halls Creek, 250 kilometres away, in search of a spare.

We arrived in Halls Creek on New Year's Eve, and found everyone busily preparing for the big dance at the town hall. It was a dour affair—all white—with the hall starkly lit, barely decorated and enlivened only by a country and western band plucking away in the corner. The young men who had come from miles around made it clear that they wanted no competition for the few eligible girls, so Bill and I avoided a one-sided fight by retiring early.

The next morning I discovered that in all of Halls Creek there was

not a motorcycle tyre that fitted my Guzzi. I would have to order one
in from Darwin, which meant further delay. To safeguard my bike I
would also have to truck it to Halls Creek, but with the wet season
setting in this was not going to be easy. I began by calling the police
station in Fitzroy Crossing, inquiring whether they could help me locate
a truckie willing to load the bike. 'You mean that Italian bike up the
track?' the policeman who answered the phone said. 'Too late, mate. It's
burned to a crisp.' When I asked how it happened, he replied, 'Abos,
mate. Would have burned it just for the clothes in the panniers.' I was
still numb when I told the garage owner I would not be needing a tyre
after all, and when I explained why he replied, 'Yeh, that cop at Fitzroy'll
really teach 'em a lesson, take his shotgun down to the camp. Least you'll
get yer clothes back.' And so I didn't pursue the matter. Instead I went
with Bill to Darwin from where, like any teenager, I phoned Mum and
asked for money for an airfare home. The city had been hit by Cyclone
Tracy the year before, and most of it was little more than a forest of concrete house supports. I felt about that flat too.

I still don't know what happened to my bike. And it was only much
later that I learned who those people were camped around the hotel at
Fitzroy Crossing. Many were Walmatjarri tribesmen and women from
Noonkanbah station. From the time their land had been appropriated
by Europeans in the late nineteenth century until the 1960s, the
Walmatjarri had worked as cattlemen, being paid by the owners partly
in flour and sugar. The life they lived was not so different from the one
they had known before the European invasion, for they were still living
on their traditional land, visiting its sacred sites and eating plenty of bush
tucker. When we granted them equal wages in the 1960s, the station
owner couldn't afford to pay union rates, so he kicked them off the place.
They had retreated to the only refuge they knew—Fitzroy Crossing.

In 1976—a year after I heard their sad corroboree—the Walmat-jarri got their land back, the transfer of title marked by the passing of a handful of soil to a tribal elder. It was one of the earliest acts of reconciliation to take place in Australia, but still the Walmatjarri had a long fight ahead of them. The mineral exploration company CRA wished to explore for oil in the area—mineral exploration was not allowed, while looking for oil was. After several confrontations between the Walmatjarri and CRA, in 1981 a hole was drilled right next to a sacred site. Not surprisingly, no oil was ever discovered, and having established the primacy of their right to explore, the mining company moved on, leaving a punctured land and people in its wake.

By the time I returned to Melbourne in the summer of 1976 I had dis-covered that I lived in the most astonishing, harsh and beautiful country on Earth; and that the kangaroo, in all its enigma and mystery, lay at the heart of Australia. I still had several years of my humanities degree before me, but now I knew what I must do. Somehow I would become a biologist so that I could better understand this country and its extraor-dinary creatures. It was a tortuous career path, involving many catch-up courses, a Master of Science in geology at Monash University, then a shift to zoology and the University of New South Wales, where I com-pleted a doctorate in 1984. By a stroke of great good fortune I then found a job—in November 1984—as technical officer in the mammal section of the Australian Museum in Sydney. Now I could study kan-garoos to my heart's content.

6

Kangaroo Essence

When the gentlemen of Hobart met in 1829 to form Australia's first scientific society, they took as their motto *Quocunque aspicias hic paradoxus erit*. 'Whatever one examines here will seem a paradox.' It was a phrase that well captured the received British perception that everything in Australia was, *contra naturam*, novel to the point of being ridiculous. The kangaroo, as symbol of the newly discovered land, bore the full brunt of this preconception. Phrases such as 'kangaroo court'— a court where there is no justice—entered the language via America and reinforced the sense that things get turned on their head in Australia and can't be taken seriously.

For anyone who has watched a red kangaroo with her joey in the dawn light of the Australian desert, such glosses are manifestly inadequate. The crisp air lends a gossamer-thin softness to the landscape whose pastel shades are all the more precious for the knowledge that in an hour or so they will be gone, and in its stillness wafts the delicate scent of dust and saltbush that is the essence of the outback. Not

A whimsical artistic rendering of the first kangaroo, as it was imagined by Rudyard Kipling.

knowing that you are there she rests, the mother, her eyes half closed as the first weak rays of the sun warm her, while her offspring tries out his new legs in flailing investigations of every bush, insect and stone in his expanded world. They are frail living things in an awesome wideness of environment that, like the open ocean, offers no refuge from the forces of nature. Yet they will survive. They always have. Unless of course we disrupt the subtle web of relationships that life is attuned to in this country.

So what, in essence, are kangaroos? While manners may make the man, the outer form does not make the kangaroo, for members of this family are astonishingly varied—being mistaken for cats, rats, deer and raccoons by early European observers. Their variety continues to confound non-expert observers, and yet no matter whether they be tiny rat-kangaroos or the giant red, kangaroos share a suite of characteristics that make them utterly different from all other living things. These kangaroo essentials, so to speak, underpin the creature's great success, and involve the way kangaroos reproduce, eat and get about. Because they evolved with the earliest kangaroos, they tell us a lot about the ecological challenges that faced the first kangaroos in a long-vanished Australia. In effect they form a sort of family blueprint, an 'anatomical fossil' dating to the Eocene period, around 35 to 55 million years ago.

Kangaroos have no close relatives among living marsupials, yet their anatomy bears the indelible stamp of having evolved from small, tree-dwelling ancestors. One of the first people to speculate on what that original creature might have looked like was Rudyard Kipling, who in 1908 published the 'Sing-song of Old Man Kangaroo':

> Not always was the kangaroo as now we do behold him, but a Different Animal with four short legs. He was grey and he was woolly, and his pride was inordinate: he danced on an outcrop in the middle of Australia…saying, 'Make me different from all other animals; make me popular and wonderfully run-after by five this afternoon.'

When biologists rather than poets try to picture the creature kangaroos evolved from, the living pygmy possums—mouse-sized omnivores that lurk in dense scrubs in southern and eastern Australia—offer the best guide, for they share a few odd characteristics with kangaroos. Hopping, however, is not one of them, and the ankles of these possums make one realise how extraordinary it is that kangaroos—and hopping—ever evolved at all. Possums have the most flexible ankles of any living mammal, allowing the foot to be 'dislocated' so that it can swing through 180 degrees and point backwards—something unimaginable outside the

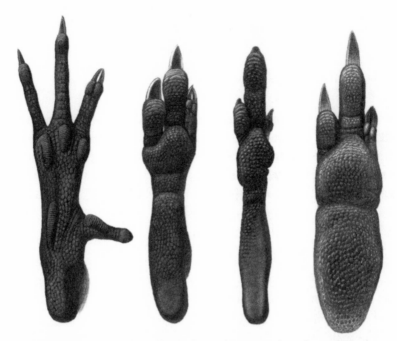

From left to right: 1) The foot of a musky rat-kangaroo, the only species to have a grasping great toe. 2) A forest wallaby. In all kangaroos two toes (here, those on the right) are encased together in skin for grooming. 3) A rock-wallaby, whose short claws and fingerprint-like gripping pads are useful when leaping among rocks. 4) The tree-kangaroo's broad foot and long, curved claws help grip tree-trunks.

torture chamber for humans and kangaroos alike—but a very necessary ability for an animal that climbs down tree-trunks head-first. To convert this most flexible of ankles into the rigid structure required to withstand the extreme forces generated by hopping is an engineering feat akin to converting a bicycle into a bulldozer.

The flexibility in a possum's foot is facilitated by the shape of its joints, which are rounded—rather like ball-and-socket structures—and loose-fitting. The joints in the kangaroo foot are, in contrast, all sharp angles and closely opposed linear facets. In extreme cases bones have become fused together to exclude all movement except in a forwards–backwards direction. The evolution of the kangaroo foot is so distinctive that a single bone is usually all you need to identify its owner. Indeed the kangaroo family, Macropodidae (meaning 'big foot') takes its name from this part of the anatomy.

We have little idea how long it took for this radical re-engineering to be effected, but we do know one thing—the changes were not driven by the requirements of hopping, but rather a shift from an arboreal to a terrestrial lifestyle. This is because it is evident from the fossil record that hopping was not achieved until long after the distinctive features of the kangaroo foot first appeared. For millions of years—possibly tens of millions of years—after the group first arose, it seems all kangaroos got about by bunny-hopping.

Feet are important, for unless one belongs to a shoe-wearing species they provide the interface between a land mammal and its environment, and that determines where one can go, how fast, and at what expense. Another anatomical feature of the kangaroo every bit as odd as its feet are its genitals. Contrary to any sense of good use—and as noted by many early settlers who took this as further proof that everything was paradoxical in Australia—in the male the usual order of appearances is reversed—the scrotum hangs in front of the penis. Unless you have seen an excited male kangaroo you may not have noticed this remarkable fact, for most of the time its penis lies hidden inside a pouch in the cloaca (appropriately meaning 'sewer' in Latin, for it is the common opening for both urine and faeces). The kangaroo penis is slender, S-shaped, pink, and very long. It needs to be, for to be of use it must snake around the prominent scrotum.

All marsupials share this strange anatomy, and if you have the necessary equipment and are so inclined, palpation of your own wedding tackle may convince you that your ancestors possessed a similarly inconvenient arrangement. To prove that the reverse order is the primitive condition for mammals, trace your penis posteriorly until you find its base, very near the anus (a more specialised offshoot of the cloaca). The ensheathing of the organ in skin and tendons so that it exits the body well forward of this point has been the work of untold millions of years of evolution.

The original, forward position of the testes in our ancestors can also be demonstrated by our anatomy. Gently pinch your abdominal wall just to the sides and in front of your penis's point of entry into the world. You should be able to feel the cords from which the testes are suspended. Before you were born your testes migrated through your abdominal wall then took a downward turn to enter the scrotum, there to reside in sensible placental manner behind the business end of the penis. But in our distant ancestors, which lived before this migration occurred, they clung to the stomach wall below the belly button. This is pretty much what you see in kangaroos, except that they have evolved a capacious scrotum to house them.

Do not, incidentally, believe everything you see in museums, for sometimes confused taxidermists have effected a kind of evolutionary sex-change operation upon the unfortunate marsupials in their care. The most spectacular example I know of resides in Oxford University's zoology museum—the very place where Thomas Huxley took on Bishop Wilberforce in the first public and fiery debate on evolution. When mounting their stuffed male thylacine, a deluded don (or a don's assistant) fabricated a whole new three-piece set for the animal, modelled precisely upon the genitals of an alsatian and crafted (as far as I can tell) out of deer fur! If you wish to see real male thylacine genitals travel to Leiden in the Netherlands, in whose Naturalis Museum resides the only surviving thylacine penis I know of. It is a splendid organ—albeit like that of the kangaroos a little on the slender side, but just as long, and double-headed (or at least cleft) to boot.

While the reproductive anatomy of the male kangaroo may be mostly a reflection of the ancestral mammalian condition, that of the female is packed with novelty. Let me give you a riddle. What has two vaginas, yet gives birth through neither? If your answer is a female kangaroo, congratulations, you are correct. Her two vaginas, which open into her cloaca, are a relic of a time when the entire female reproductive system was divided into left and right sides, a condition that can still be seen in reptiles. But the fact that kangaroos do not give birth through either vagina is entirely original. Instead, as her time nears, a canal forms in the flesh between her vaginas, and it is through this unique anatomical structure that the young enters the world.

Like all mammals the young kangaroo begins life as a fertilised egg. At one tenth of a millimetre in diameter it is tiny, but as with the eggs of chickens (and unlike our own eggs, which lack the outer layers) it is enclosed in a full set of shell membranes. It is out of this tiny egg that, during the last week or so of pregnancy, the kangaroo hatches. Its egg-yolk membranes then form a sort of placenta, delivering nutrients directly from the mother's uterus, allowing the embryo to grow rapidly. After a pregnancy of thirty-three days the red kangaroo is born, having grown to the size of a raisin and weighing just under a gram. That such a minuscule being can breathe, climb and suckle is a marvel to many people, but the young kangaroo is born in an advanced state compared with the young of other marsupials, which have shorter pregnancies and smaller young, even when adult body size is taken into account. Some bandicoots are pregnant for only eleven days, while many marsupial newborns are the size of a grain of rice.

As befits a species that, relative to other marsupials, invests so much in pregnancy, the mother kangaroo lavishes resources on young that are born singly or, rarely, as twins. Again this contrasts mightily with the basic marsupial pattern, of which the American opossum is a fine example. An opossum mother has thirteen nipples—the one making up the baker's dozen sitting in the middle of a circle created by the rest—but she gives birth to as many as fifty-six young, thus dooming most to lose at the first lotto-draw of life. The production of throwaway embryos in this manner is common among marsupials, but is utterly foreign to the kangaroo.

The kangaroo possesses a remarkable (at least from a human perspective) capacity to switch on and off the development of her growing young. Were women so endowed they could conceive whenever they wished, then tailor the baby's birthdate to suit their calendars. While there is still much mystery about the origins of this amazing capacity, kangaroos use the ability to maximise their reproduction. Once she reaches sexual maturity the female kangaroo is, quite literally, forever barefoot and pregnant.

The system works as follows. A few hours after the female kangaroo gives birth and houses her new baby joey in her pouch, she seeks a mate, copulates and once again falls pregnant. Having sex on the birthing bed, so to speak, might seem to be in bad taste and contrary to common sense—so why would any creature do it? The answer lies at least in part in energy efficiency, for pregnant and lactating mammals must maintain a higher body temperature than normal, which costs extra food and energy. If you can be pregnant and lactate at the same time, you shorten the length of time an elevated temperature is required, thus saving on energy.

Following conception the embryo develops normally until it is about a third of a millimetre across, then it abruptly stops. A chemical 'order' has been sent for it to cease growing, and it enters a state of suspended animation in which it can remain for almost a year. Inevitably a signal to resume development is sent, and within a few weeks—usually on the very day that the older young leaves the pouch permanently—a new kangaroo is born. Two other Australian marsupials possess a similar capacity, the thumb-sized honey possum of the southwest of Western Australia, and at least one species of pygmy possum, which is one reason why scientists posit an evolutionary link between kangaroos and these creatures. Several kangaroo species have modified this reproductive heritage—the female eastern grey kangaroo does not copulate until three or four months after the birth of her young—but the pattern outlined above seems to have evolved with the earliest kangaroo, and it is retained in the majority of her descendants today.

Interestingly, the puppet-master controlling the development of the embryo is not the mother at all, but her joey. While it sucks hard and steadily, the embryo will remain in suspended animation. But if the suckling eases—perhaps an accident befalls the joey or it starts to eat solid food—the quiescent embryo recommences its growth. And while the older young is shut out of the pouch by the mother's powerful pouch muscles on the day of its sibling's birth, the mother does permit the older joey to put its head into the pouch to suckle from its nipple (joeys remain faithful to a single nipple, despite the fact that they have four to choose from) which after many months of sucking has become enormously elongated.

The mother must now suckle two young of very different sizes—one weighing a gram and the other up to four kilograms—and because their nutritional needs are very different she must provide milk with differing chemical formulas. To her smaller young she must provide the proteins needed to grow fur when it is ready, while the larger young may need more water. Just how she performs this miracle of mammary discretion no one understands, and few fields in the natural sciences offer more promise of fundamental discovery than that of marsupial lactation.

Taken as a whole this astonishing system represents a reproductive 'conveyor belt' which is poised, forever ready with another 'product' to roll off at the earliest opportunity. It is perhaps the major factor in the success of the kangaroos, but there are other anatomical features that have also helped them triumph.

The final 'kangaroo essentials' relate to diet, and are manifest in the teeth, jaws and stomach. This is fortunate for palaeontologists because most vertebrate fossils are teeth, which permits easy identification of kangaroos in the fossil record. Kangaroo jawbones are perforated by a large hole, through which a cheek muscle enters to anchor far forward and deep inside the bone. This allows great force to be delivered in a part of the jaw which in primitive kangaroo species is occupied by a blade-like and finely ridged premolar. The entire system appears to have evolved to deal with food that needs considerable force to break into, but which is soft inside. The serrated premolars have been lost in many kangaroo lineages, but the hole in the jaw invariably remains, and is a sure guide that the bone belongs to a kangaroo.

The food that those premolars broke up then made its way to a rather odd stomach. All living kangaroos possess large stomachs in which different 'sections' can be discerned (albeit less distinctly in more primitive species). This compartmentalisation was to become a most important development in those lineages of kangaroos that evolved to eat grass, for it permitted a highly efficient form of digestion.

Because the specialisations listed above are common to all kangaroos, or somewhat modified in a few lineages, they must have been present in a common ancestor. Using them we can reconstruct some of the challenges and opportunities that this creature, which evolved over 30 million years ago and for which we have almost no fossil evidence, must have faced. They indicate that the ancestral kangaroo was spending more time on the ground. Was this because forests were vanishing, or because new opportunities were opening up at ground level? Perhaps it was a bit of both, but the opportunities on the ground surely included some foods difficult to break into. For an animal not much larger than a pygmy possum, many foodstuffs can be hard to crack: seeds, nuts, insects with tough shells and even snails are all possible candidates.

And the unique method of kangaroo reproduction suggests that there were threats in this new environment as well as opportunities. You can tell a lot about the risks an animal faces from its method of reproduction. Species with high predation rates typically have many young, while those that face fewer predators tend to invest more in raising fewer offspring. Because the ancestral kangaroo bore a single young upon which it lavished much care, we can infer that once the perils of weaning were passed the joey had a fair chance of reaching old age. For an omnivorous creature the size of a rat this is rather surprising, as there are plenty of predators capable of killing such an animal in most ecosystems.

But what of the delayed development of the embryo? Some scientists believe that in the early stages of its evolution the process served to extend pregnancy (and thus care of the young), and to increase energy efficiency (by allowing the mother to be pregnant when lactation was at its height, thus minimising the time needed to maintain a high body temperature). Nevertheless, by the time the ancestor of the living kangaroos had evolved, the process was probably reducing delays between births, and into that we can read several evolutionary meanings. Having an embryo waiting, quiescent in the uterus, is useful to animals that stand a fair chance of losing their young before weaning, for it allows them to rapidly replace any losses. Despite the rare occasion upon which a mother ejects an immature young from her pouch, most pouch-young are safe from being eaten as long as their mother survives. Starvation is the greatest killer, for young, growing bodies demand food of far higher quality than that required by adults. Today, moderate droughts will kill innumerable young kangaroos, which raises the possibility that drought may have been a feature of the Australian environment at least 30 million years ago, when the first kangaroos appeared. If upheld by further research this is an important discovery, for many of Australia's plants and animals tolerate drought, yet enigmatically there has been little evidence of drought in the fossil record before around 8 million years ago.

The delayed development of the embryo implies one further feature of the environment that shaped the ancestral kangaroo: it was not strongly seasonal, for to state the obvious, species in seasonal environments give birth seasonally, not when an older sibling ceases suckling. Furthermore, those few kangaroos that breed seasonally today have had to modify the original pattern. The absence of strong seasonality is interesting when one considers that, before 30 million years ago, Australia was located far south of its present position. Indeed, until around 40 million years ago it was still attached to Antarctica and a part of the ancient supercontinent of Gondwana.

So, was the first kangaroo grey, woolly and with 'pride inordinate' as Kipling would have it, and did it dance upon an outcrop in central Australia? We still do not know when and where kangaroos first arose, but we can look to modern environments that have some of the features that our analysis suggests the first kangaroos experienced. The high mountains of New Guinea, with their frigid, mossy forests, offer a good starting point. Firstly, they are composed of ancient Gondwanan plant species of a type that we know existed in Australia in times past, and many of these bear hard seeds and nuts. Secondly, these forests have very few predators capable of eating a large rat-sized creature. Only the New Guinea quoll and a few owls fall into this category, for the environment is far too cold for large reptiles.

At altitude, towards the tree-line, there are also open spaces where tree-dwelling creatures will come to ground—indeed, during field work in the region I have caught many a tree-dweller taking a nocturnal wander across the grassland. Only the droughts seem to be lacking, but El Niño brings periods of dryness and frost—if not absolute drought— even to New Guinea's wettest mountains. The ancient Australia that gave birth to kangaroos may have been as cool and predator-free, but it was surely far flatter, and its open spaces more likely to have resulted from infertile soil or rocks. Thus, although we can say nothing definitive about the colour or pride of the first kangaroo, perhaps Kipling was correct in placing him upon an 'outcrop'. And he is far more likely to have danced than hopped.

Trying to determine just when and where the first kangaroos lived has been an obsession throughout my professional career. In 1981, when I began my doctoral studies on kangaroo evolution, so little was known that searching for kangaroo origins in the vastness of Australian evolutionary time was akin to looking for a needle in a haystack. Then we had no idea even of when the first marsupials arrived in Australia. Was it over 100 million years ago, when marsupials first show up in the fossil record in Asia and North America, or after 65 million years ago, when they first appeared in South America?

Back then there was but one hypothesis to guide me—an estimate published by molecular biologists that all kangaroos whose DNA they had examined shared a common ancestor 50 million years ago. But this estimate proved to be dubious because it was based upon the rate at which kangaroo DNA changes, and in order to establish that rate well-dated fossils were needed to calibrate the molecular 'clock'. As such fossils were lacking, it was a classic catch-22. Based on my construction of the ancestral kangaroo, two possibilities seemed worth investigating. Perhaps the lineage was a very ancient one and its first member had come to live on the ground in the wake of the extinction of the dinosaurs 65 million years ago. I thought this was possible because those extinctions had carried off so many creatures that a multitude of ecological niches were left vacant, and perhaps an opening at ground level was sufficient to induce one of the surviving marsupials to forsake the trees for life on the ground. Or perhaps the ancestral kangaroo came to ground 55 to 33 million years ago, when worldwide changes in the Earth's climate broke up the rainforests that had dominated the planet for millions of years, creating a new evolutionary frontier in the form of open woodlands and plains. With these two possible explanations in mind, I began searching through fossil deposits from 120 million to 10 million years in age, looking for evidence of early kangaroos.

7

Dead-end in the Inland Sea

One of my first opportunities for field work came courtesy of my friend and mentor Dr Tom Rich. One day in 1977, while I was cleaning fossil bones in his lab, Tom let slip that a team of scientists from the British Museum of Natural History was coming to western Queensland. They intended to spend two months searching for dinosaurs in sediments that had formed along the margins of Australia's inland sea over 100 million years ago. Tom was going to join the group because he thought that the region might yield Mesozoic (dinosaur age) mammal fossils.

Back then, the record of fossil mammals in Australia was, as Tom put it, a truly ghastly blank, without a single mammal bone having been discovered that was older than about 30 million years. Quite frankly, many of us thought Tom was chasing a phantom, for we felt that such mammals may never have existed in Australia.

It was to be an epic quest. Australia's federal research funding body had other priorities, so Tom turned to the land of his birth, and more specifically to the National Geographic Society, for aid. The fossils he was chasing proved exceptionally elusive, yet for over two decades the society kept the faith, during which time their dogged palaeontologist found so many dinosaur bones that he became a world authority on the Dinosauria; and such an abundance of fossils from the later Cainozoic era (the age of mammals) that Tom is considered an expert on these as well.

When he asked if I would like to join him in Queensland I felt as if I'd died and gone to heaven, especially when I learned that among the expeditioners were some of the gods of my teenage reading: people like Alan Charig, whose book on dinosaurs entranced me, and Barry Cox, whose publications on fossil mammals I had spent hours puzzling over. The chance to meet such scientists and to look for dinosaurs in western Queensland seemed too good to be true. And what exciting rocks! They had their origins in deltas, swamps and floodplains that had surrounded Australia's inland sea. Between 120 and 100 million years ago this long-defunct ocean had occupied much of what is now inland Australia. Then, Uluru (Ayers Rock) and Kata Tjuta (the Olgas) must have stood as lonely sentinels in an expanse of azure, home perhaps to nesting pterodactyls, turtles and Loch Ness monster-like plesiosaurs, while along the eastern margins of the sea, great rivers coursed through the Australian uplands, laying down the rocks that we were to visit.

The stories of Aborigines and the westward flow of many rivers convinced Australia's pioneers that an inland sea existed, prompting many an explorer to set out in search of it. Some, such as Charles Sturt, even dragged boats over the plains, turning an arduous venture into a total nightmare. But now, we were off to prospect for evidence of life from around that sea—an adventure only slightly less quixotic, perhaps, than that of the ill-fated explorers.

The British had been in the field for several weeks by the time Tom and I arrived at the dusty outback town of Winton. The place resembled the set of a Wild West movie: endless plains, wide verandahs and the lure of dinosaurs induced a state of high excitement in my young heart. Our rendezvous was half a day's drive down the road in the big sky country out towards Birdsville, through a landscape dotted with cattle, and every now and then a barbed-wire fence festooned with the carcasses of eagles and hawks, testimony to a hatred of predators then felt by so many rural people. To me this was a revolting sight, for the graziers had hunted down the very animals that were their best allies in the war on rabbits. Yet there they were, from the rarest to the most common, and in varying states of decomposition, laid out with military precision, victims of a blind hatred that had come with the first European settlers to this land.

Southwest of Winton the Diamantina River has cut a wide valley filled with bountiful black-soil plains that in a good season are waist-high with grass. To the east they are bounded by a plateau around 200 metres high known locally as the 'jump-up'. Its margin has been eaten into by creeks and gullies for millennia, forming a rough jumble of bluffs, mesas and narrow valleys whose white, yellow and pink sandstones and chalks give it a distinct 'badlands' feel. These were the rocks we had come to investigate and I hoped that somewhere out there, on a wide wash-out or nestled amid the tufty grass were dinosaur and mammal bones—maybe even entire skeletons. The British, however, had other ideas, for they began their survey not on the plains but in outback pubs.

Alan Charig turned out to be a boyish man with a wild laugh and a wicked sense of humour, whose bright blue eyes, peering out between a haphazard straw hat and an unruly black beard, gave him the appearance of a pirate of the sort found in *Treasure Island*. I immediately liked him, and was eager to learn from him, but when he explained his team's strategy to find dinosaurs, I thought he was pulling my leg.

'Queensland cattle-men,' he explained, 'are a curious and thirsty people. While out mustering cattle they must come across plenty of fossils which they occasionally bring into their pub to show their friends.' Charig reasoned that these fossils sometimes get propped up behind the bar, at least until a palaeontologist wanders by. And he proved to be spectacularly correct, for in a couple of weeks of pub-crawling the British turned up more fossils than we would see in a month's intensive on-the-ground survey. One of their most remarkable discoveries had been made while enjoying a cool beer in the airconditioned comfort of the Waltzing Matilda Hotel in Winton. A member of the team spied a particularly fine bone from an armoured dinosaur propping open a door. If it turned out to be a new species, he explained, the British would call it *Matildasaurus* in honour of its place and method of discovery.

Charlie, the owner of the station we used as base camp, had agreed to allow our expedition to occupy an outbuilding. We would rise at first light and drive to where we had left off the day before. It was necessary to cross country to do this, and we would frequently encounter low-slung 'party line' phone wires, then the only means of telecommunication for many rural Australians. While they offered little in the way of privacy (anyone picking up a phone could listen in on the conversations), they were a vital lifeline in this remote country. The jerry-rigged wires were strung on crooked old mulga posts just high enough for a Holden ute to pass under. When we misjudged their height or just did not see them our Land Rover would snag the wire on a roof rack, perhaps cutting a conversation mid-sentence. Then we'd spend hours repairing the break, stretching the wire back into place, all the while hoping that we had not interrupted an emergency call and that the posts would hold.

Having arrived at the outcrop we would fan out and walk over the stark countryside, converging at a predetermined pick-up point in the afternoon to return to base camp before dusk. Although dinosaur bones were scarce, the region fascinated me, for you never knew what lay around the corner. One spectacular fossil site was already known. In the late 1960s a local, Ron McKenzie, had come across some strange marks in a purplish-pink rock. They were the tracks of chicken-sized dinosaurs—thousands of them. When the layer containing them was excavated it was discovered that a massive carnivorous dinosaur, whose tracks were also beautifully preserved, had scattered a herd of small herbivores. The tracks had been formed in mud beside a waterhole, and probably extended over several hectares, only a small part of which had been unearthed.

Charlie told us that 'black wallabies' had once inhabited the jump-up. They may have been rock-wallabies, but along with all of the smaller members of the kangaroo family they had vanished from the region at least a generation earlier. All we saw in the hills were roan-coloured euros. They were in abundance, not because the country had been over-grazed, but because these rocky ranges were an ideal habitat. Caves were full of their droppings, and their footprints through the rocky pinnacles often served as my pathway. Sometimes, in the early morning or afternoon, I would spy the silhouette of a euro resting under an overhang. If I approached quietly I could get quite close before the creature bounded away, breaking the silence of the bush with the clatter of dislodged stones. Occasionally I would surprise groups—a mother and young, and perhaps a male waiting nearby as well. Wandering alone in such country, day after day, was a dizzying experience for this twenty-year-old, for at last I was exploring the wild outback. After hours of absorbed walking with my eyes fixed on the ground I would look up to find myself surrounded by a labyrinth of bluffs and mesas and realise I had no idea where I was. The temptation was to panic, but if I could suppress the urge and sit quietly on a high spot, the sun and the lie of the land would invariably give some direction to the rendezvous point.

Truly unexpected things occasionally burst upon us. A furious squeal and wild rushing through waist-high grass would indicate that we had roused a sounder of feral pigs from their midday rest. It took some nerve to stand still as the unseen creatures, the males armed with razor-sharp tusks, rushed by our legs. Once, on the rubble-strewn shoulder of a mesa, a glint of deep-ocean blue caught my eye. I picked up the cobble, spat on it and rubbed away the dirt with the tail of my shirt, revealing a double seam of opal glistening like a vein of tropical seawater in the red sandstone—all the more glorious for the desolate surroundings.

One morning we located a canyon leading deep into the jump-up. The sunlight was all but cut off by the steep walls, and lusher vegetation grew along a sandy wash-out with traces of moisture at its base. As we penetrated further the water began to flow, then figs appeared perched on a rocky precipice high above. Finally the gully narrowed, becoming dark and ferny. On the higher walls was Aboriginal art—hand stencils, boomerangs, kangaroos—and under the deepest overhangs lay bundles of paperbark. The remnants of Aboriginal burials in bark coffins. Why, I wondered, had I not seen any Aborigines around Winton? Western Queensland was, it transpired, a particularly bleak place to be an Aborigine in the nineteenth century with one squatter, C. B. Dutton, summing up the situation in a phrase: 'You are black, and you must be shot.' No one there could tell me anything much about the local tribes.

After weeks of searching Tom found our first fossil, a bone reduced by wind and weather to a pile of rubble, its fragments scattered over a hundred square metres of wash-out. We diligently gathered up the splinters and spent the evenings piecing them together so that by the time we left we had assembled most of the shoulder blade of a huge sauropod—one of the long-necked, plant-eating dinosaurs.

We had been eating out of tins for weeks when the team made a deputation to Charlie to sell us some meat. His shaded eyes peered at us from under his broad-brimmed hat as he laconically asked, 'What cuts do you want?' Tom asked for fillet, but when Alan Charig, with his impeccable English accent, said that a bit of liver might be nice, the slightest smile flickered across Charlie's weatherbeaten face. He turned to me and to the youngest Pom, David Norman. 'You two fellas come with me.'

I was relegated to the back of the utility as we bounced towards a group of grazing steers. We stopped around fifty metres from the herd and the barrel of a rifle appeared from the driver's side window. Bang!

A white-faced steer fell to the ground, blood spurting from its forehead. The gun retracted as the remaining cattle ambled off, not at all alarmed. Charlie opened the door and strode towards the fallen beast, a large knife in his hands. David was aghast, for like me he was a city boy who had eaten innumerable steaks yet never seen a steer die. But things became much worse when the animal rose unsteadily to its feet and lurched a few steps, blood pouring from the bullet-hole in its head.

Charlie ran at it from behind and expertly slit its throat before riding the dying beast into the dust. Within moments he was butchering it, its skin becoming a ground sheet to keep the cuts clean. As we helped load the meat onto the tray, Charlie passed me a long strip of scotch fillet. Then he emerged from behind the ute with the liver. The bloody, purple organ seemed a metre across. David turned a more awful shade of green as he accepted the offering and laid it in the back. When we arrived at the shed, our British colleagues went off the idea of offal, causing the barbecued fillet to become somewhat finely divided.

After this experience I began to think about meat-eating and animal rights in quite a different way. Charlie must have frequently slaughtered a beast to feed workers and family, and his botched shooting was surely a rare embarrassment. Was the fate of that steer, I wondered, any worse than that of those taken to an abattoir for slaughter? Its end was, I suspect, significantly less painful and traumatic than the slaughterhouse-bound majority, for the creature went from calm grazing to the stillness of death in a few seconds, avoiding the round-up, transportation by road and queueing before the slaughterer at an abattoir.

Would it not be morally preferable to avoid eating meat? What, then, would become of the outback, which is unsuitable for agriculture? Without industry no one would live there and manage the land, so central Australia would become a vast degraded reservoir of feral animals, in which native species and introduced ones alike would, in drought, suffer and die by the million.

Care for our ecology must underpin everything we do, for without a viable ecosystem humans and animals will not survive. And some-times, in order to stabilise the environment that we have so badly damaged, it is necessary to kill and to cause suffering. Just think of the mass death and pain of rabbits caused by myxomatosis or calicivirus. We must, of course, seek to minimise that suffering in every way we can; but we must also be willing to face the difficult decisions that are inherent in our role as the most powerful force in the environment. That is why I think people who kill their own meat, in as humane a way as possible, are the most moral of us all. In doing so they develop the understanding, courage and compassion for life that are fundamental requirements of the 'decent' person, things that those of us who receive our meat in plastic trays have little opportunity to achieve. It is as if we are inhabitants of a great feedlot—albeit an urban one—which robs us of full control over our lives, in particular our consumption of energy, water, food and material goods. Worse, it compromises our morality.

After a month of intense effort Tom and I had found only four dinosaur bones, three of which were mere fragments. Tom was particularly disappointed that we had failed to find even a rock type that looked as if it might yield fossil mammals. And the only living marsupials we saw in the area were euros and red kangaroos, the latter only glimpsed when bounding across the plains. It was around the time I was admiring them that the secret of this animal's effortless locomotion was being investigated—not on Australia's inland plains, but in a laboratory in faraway Harvard University.

8

The Mystery of Hopping

The disastrous wreck of the Dutch vessel *Batavia* on the Abrolhos Islands off Western Australia in 1629 provided the first opportunity for Europeans to observe a member of the kangaroo family. Captain Franz Pelsaert saw large numbers of creatures he described as 'cats' on the forlorn islands where the shipwrecked mariners sought refuge. But they were strange cats, for their hind legs were 'upwards of half an ell in length [about half a metre], and it walks on these only, on the flat of the heavy part of the leg, so that it does not run fast.' These creatures were tammar wallabies (*Notamacropus eugenii*), smaller relatives of the great red and grey kangaroos, and Pelsaert's appraisal of the wallaby hind-leg is quite accurate, leaving no doubt as to the identity of its owner, yet it seems that the tammars never hopped in Pelsaert's presence, or moved rapidly at all. The Dutch would surely have welcomed the five to ten kilograms of meat that each wallaby offered, so why did the creatures not flee as they do best—by hopping away? Perhaps after 10,000 years of isolation on their arid island, where the only predators were eagles, they failed to perceive the danger that the Dutch represented.

Whatever the case, had Pelsaert observed his 'cats' more closely, he would have seen something to make any shipwrecked sailor envious—the creatures can drink saltwater. This means that tammar wallabies remain fit and healthy, even reproducing when they have nothing to eat but dry food, so long as they can sip from the briny. They therefore thrive on arid isles from the Abrolhos to Kangaroo Island.

The Dutchman may have been further astonished had he known that, like horses, tammars share a common birthday, for the great majority enter the world in late January. The tammar is one of the few species of kangaroo to have modified its ancestral reproductive pattern to become a seasonal breeder. The embryos emerge from suspended animation around the summer solstice (22 December) and are born a month later, ensuring that grass greened by winter rains is available to them when they emerge from the pouch in another eight to nine months.

By the seventeenth century, Dutch mariners had recorded the existence of both quokkas and tammars, but it was not until Dutch artist Cornelis de Bruin encountered a member of the kangaroo family that the world received an accurate description. The year was around 1700, and the location exotic—a colonial garden on the island of Java. He had been invited to visit the governor-general of the Dutch East Indies at his country abode, and there observed an animal he called 'filander', a corruption from the Malay 'pelandok Arou'. Judging from de Bruin's illustration, the creature was the Aru Islands pademelon (*Thylogale brunii*) which also inhabits southern New Guinea. The transplanted colony was thriving, and enjoyed:

full freedom, running with some rabbits which have their burrows under a little hillock encircled by a fence. The Filander, which has hind-limbs much longer than the fore, is nearly the size of, and possesses nearly the same form as, a large rabbit…but the most extraordinary circumstance is that the female has a bag-like opening in the belly into which the young enter, even when they have attained a considerable size. They are often seen with head and neck thrust out of this bag; however, when the mother is running the young are not visible but keep to the bottom of the pouch.

The wallabies were breeding well in exile, for whenever the governor held a feast, his tables 'groaned under the weight of the Aroe rabbits'.

It was not until James Cook led his nation's first major scientific expedition, which in 1770 charted Australia's east coast, that a true appreciation of hopping was gained by the Europeans. But it fell to Dr Terrence Dawson to unlock the deep mysteries of hopping. In his 1995 publication *Kangaroos: The Biology of the Largest Marsupials* he writes that as a young researcher at Harvard people expected him to know things about kangaroos which neither he nor anyone else then understood. Dawson decided to carry out a thorough investigation into the seemingly obvious—how and why kangaroos hop. His main tools were a treadmill, a few red kangaroos which had been brought to Boston and trained to hop on it, and a battery of devices to measure oxygen consumption, muscle effort and heart rate in the gymnastic marsupials.

What he discovered was amazing. Hopping at medium speeds (15 to 40 kmh), Dawson and his colleagues concluded, is the most efficient means of locomotion ever evolved by a land-bound creature. Much of the energy expended in hopping is saved in the tendons of the legs, which act like the springs in pogo sticks, storing the power of each bound and releasing it to assist with the next. The heavy tail also stores energy as well as acting as a balance. Later studies by other researchers demonstrated that even more energy is saved by the action of the gut, which moves like a piston with each hop, emptying and filling the lungs and thus saving the effort of breathing.

More recent research by Dawson and his students has uncovered further remarkable aspects of the marsupial metabolism. The heart of the red kangaroo is twice the size of that of a similar sized placental mammal, such as a deer, and when at rest it beats only half as often, thus saving energy. But when at work it can beat up to 60 per cent faster, allowing for a massive sustained output when required. The tail is also prodigiously powerful, exerting as much force as it pushes a roo along at low speeds (less than 6 kmh) as both human legs do when walking.

The origin of the kangaroos was very much on my mind when, in late 1984, I got my first real job. Following my doctorate I had been appointed to the mammal section of the Australian Museum in Sydney, and a phone call from Dr Alex Ritchie, curator of fossils, soon brought an exciting discovery. 'There's something here you really must see,' he blurted out in his brogue, before giving me the address of a nearby motel. Alex opened the door to reveal a suntanned opal miner and, behind him, spread out over the bed, were hundreds of opal fossils.

In opal fossils we see nature's rubbish transformed into precious gems. No one fully understands how a shell, piece of wood or bone is turned into opal, but such fossils are found only in a few locations in New South Wales and South Australia. At these special sites, miners dig into sediments that were laid down in or beside the inland sea, and at depths of up to twenty metres they find shells, dinosaur bones and other fossils that flash with red, green and blue. The opal fossils on the bed had been brought to the museum for sale, and among the usual clamshells, fragments of turtle carapace and other bones, one stood out.

It was a jawbone that had once belonged to a creature the size of a cat, and it still bore three teeth. It was a magnificent specimen—as much a jewel as a scientific treasure, for through the flashes of opalescence that emanated from it one could see the internal structure of the bone, which seemed to have been replaced by a beautifully tinted glass. Through this the roots of the teeth and the channels that once conducted nerves and blood vessels could be seen. It was as if, after being buried, the entire bone had been delicately etched away, leaving a void into which the opal had been deposited. Yet so delicate was the process that even fine films of clay, such as those surrounding the tooth roots, were left in place as the opal formed around them.

The jaw represented a breakthrough, for it was around 110 million years old—four times as old as any mammal ever discovered in Australia. But what sort of mammal was it? I had half expected the bone to be from an ancient marsupial—perhaps a distant relative of the kangaroo—but it was in fact the fossil jaw of an ancient platypus. In 1985 the jaw, along with other opalised fossils, was purchased by the Australian Museum for $80,000. The news caused quite a stir among the opal miners at Lightning Ridge and I hoped that more fossils would be forthcoming. Within days I received a phone call from an old miner who announced in a conspiratorial whisper that he had located the complete skeleton of a ten-metre-long dinosaur on his claim. Such a specimen, if preserved in opal, would be one of the most important fossils ever found, and would be worth millions of dollars. I was thrilled at this news and was mentally making arrangements for an impromptu trip out west when he said, 'Yep, I've outlined him perfectly on the surface with stones.' After some probing it emerged that the 'skeleton' was still buried fifteen metres underground, and divining its presence had been quite a business. The caller told me that he had been left in charge of his son's electrical store, who had urgent business interstate. This unprecedented opportunity had allowed the old tinkerer to construct a fossil-detecting apparatus consisting of a large head-frame surrounded by magnets and an electrical current supplied by an array of batteries. The maker of this outlandish device had wandered the scrub for days before making his grand discovery.

Some scepticism must have crept into my voice because the caller suddenly volunteered, 'I can find out anything with this machine, you know. I can tell yer how much money yer have in yer wallet right now!'

Seizing on this assertion, and remembering that I had not been to the bank, I asked, 'All right, how much is in it then?'

After several minutes fumbling he replied, 'Son, the signal's too weak from here. Better if you come out and look at the dinosaur, and then I'll tell yer.'

A decade later, in 1994, another opalised mammal jaw was unearthed in the area, providing a plausible excuse for a visit to Lightning Ridge. This second jaw, which was also around 110 million years old, had molars that resembled miniature hot-cross buns. After exhaustive comparisons I concluded that it too had belonged to a platypus-like creature, but one adapted to eating hard-shelled food such as clams. Perhaps it was Australia's answer to the sea otter, although pre-dating that creature by 100 million years.

The specimen had been found by a schoolteacher who spent his spare time chasing opal underground. He had been careful to recover any fossils he found during his work, and was happy for me to visit him. His camp was an eye-opener to the miner's way of life, consisting of no more than a few swags and utensils around a large fire. Underground, though, was a different story. The shaft leading to the opal-bearing layer was only a metre wide, and a long series of rusting, linked ladders, disappearing into the gloom, hung from one side. After descending around twenty metres the shaft opened out into a spacious cavern where a digging machine, blower and electrical cables lay.

All around the walls I could see mud from the margin of the inland sea, with ripple marks and tiny channels still intact, and here and there the dull glint of opal. This was where the fossil had been unearthed. The teacher had spotted it in the wall of the mine when he and another man were clearing away clay with a jackhammer. The teacher had shouted for his mate to turn off the hammer to avoid damaging the specimen: his mate refused. He was there to dig opal, not fossils, he said, as he pointed the hammer at the priceless relic. After a brief tussle the teacher grabbed the specimen from the wall, snapping off both ends in the process. As a result we may never know if the creature had a bill like a living platypus, or just how the jaw articulated with the skull.

For every precious fossil recovered at Lightning Ridge ten thousand must be lost to the digging machines and the tumblers that wash the opal dirt, in the process rolling priceless fossils of unknown creatures to nubbins.

By 2001 my hopes of finding dinosaur-age ancestors of the kangaroos had all but vanished. So when a chance arose to pierce to the heart of the inland sea I travelled with different motives. I was by this time director of the South Australian Museum in Adelaide and was preparing for a new exhibition of opalised fossils. The mayor of Coober Pedy, the leading opal-mining town in Australia, some 1300 kilometres west of Lightning Ridge, had invited me to participate in their inaugural opal festival. It was an opportunity too good to miss as miners from across Australia were expected to attend.

The drive from Adelaide to Coober Pedy takes about nine hours and crosses a stark transect from Australia's green fringe to its dead heart. The stately river red gums and winter-green croplands drop away after Port Augusta, then mulga gives way to a scattering of saltbush. As one approaches Coober Pedy even this thin cover exhausts itself and all that is left is a panorama of broken rocks and blowing dust known as the Moon Plain. Once the floor of the inland sea, it is still littered with the debris of that past age and is a Mecca for palaeontologists. Coober Pedy supposedly means 'white fellow in a hole', and most residents live underground—the only sensible thing to do in such a place. The quality of the dwellings varies enormously; some are palatial, while others are but a drive in the side of a hill whose entrance is draped with a piece of burlap, which when swept aside reveals a dusty swag and a caravan stove. Some of the more basic homes are located far out in the distant opal fields, and when enjoying the hospitality of a miner living in such circumstances, I often wonder how they survive through the summer when for weeks on end the thermometer refuses to dive below the old century Fahrenheit mark. Fate can play cruel tricks on such men. One resident showed me where he had discovered $20,000 worth of opalised shells when enlarging his bedroom. He had been sleeping for decades with his head just inches away from the cache, often desperate for a dollar or two for food.

I had travelled to Coober Pedy with Ben Kear, a doctoral student who was studying the reptilian giants that thrived in Australia's ancient inland sea. He told me the stones that lay strewn so thick across the Moon Plain had come from far and wide—some as far afield as Cobar, while others hailed from the Gawler Ranges and Broken Hill, hundreds of kilometres to the south and east. How had they got here? Some, it transpired, bore unmistakable evidence of transport by ice. The inland sea that in my imagination was a tropical paradise, was at times a field of icebergs spawned from great glaciers that ringed its southern margin. One hundred and twenty million years ago Coober Pedy was almost over the South Pole—a strange fate you might think for a place destined to lie at the baking heart of Australia.

As part of the festival, Ben and I had offered to conduct a fossil identification workshop for miners. Most specimens brought in were of limited palaeontological interest—opalised driftwood mostly, from the stunted forests that grew around the inland sea. A few very important specimens, however, did turn up, among them a magnificent vertebral column of an ichthyosaur preserved in precious opal. The specimen was famous across Australia's opal fields, and I was delighted when the miner offered to lend it to the South Australian Museum. He said that ever since he had migrated from Croatia many years ago, Australia had been good to him and he wanted to give something back.

It became obvious that some of the miners had a 'colourful' past. Many were known only by their Christian name or nickname, and with the coming of the goods and services tax (GST) they were feeling uncomfortable. A representative of the Australian Taxation Office was present at the festival to explain this new tax. To a packed house he stated that, by virtue of their driver's licence, passport or vehicle registration, they were all known to the government. It was no use trying to hide, for the taxman would find them, wherever they were. In response to this chilling news one opal buyer stood and said that he supposed it was all right to record his customers by the names they gave him, for he had purchased $200,000 worth of opal from Adolf Hitler last week, and $250,000 from Attila the Hun the week before that. The tax man was not fazed, replying simply that where 'reasonable suspicion existed' the buyer was duty-bound to record Mr Hitler's car registration number and pass it on to the tax office. A sustained silence reigned thereafter, during which an old miner sitting next to me whispered, 'There's many an uncapped shaft around here. That bastard will be lucky to get out of town alive!'

The next day an elderly miner arrived at our workshop. From the tips of his shoes to the end of his long scraggly beard he was stained the orange colour of desert dust, and he glanced around conspiratorially before delving into his pocket to retrieve a small vial of water, such as miners keep their treasures in. 'What do you make of this?' he said as he slipped it into my hand. Peer as I might I could make out nothing in the liquid but a few specks of dust. After a prolonged silence he said, very deliberately, 'It's an opalised worm jaw, and the last time I showed it to anyone they offered me five thousand dollars for it. How much will the museum pay?' The old fellow looked excitable, and I feared making an insulting offer.

'Well, it's a fascinating specimen,' I said, sucking my teeth and buying time. 'The opal value must be quite extraordinary—five thousand dollars at least—but our interest is fossils, and the fossil value of this particular worm's jaw is limited.'

Without a moment's hesitation the miner snatched back his precious vial and, glaring at me, slunk away.

My only official duty was to greet the opal festival parade as it arrived at the showground, and kiss the opal queen. My first glimpse of that august person came at 8 am—the mercury rising sharply—atop an enormous flatbed truck which ground its way at walking pace up the main street towards the showgrounds. She was a lady of generous proportions, resplendent in fishnet stockings, very short miniskirt, low-cut blouse and silver tiara. Her throne was an aged couch short on stuffing, and exposing several springs. Then came the drag-racing fraternity, whose display consisted of a defunct dragster that had been set alight, which was towed by an only slightly less disreputable vehicle. The fire brigade followed, playing their hoses wherever things seemed to be getting too hot. Most of the town brought up the rear, led by 'Bad and Ugly' a pair of identically dressed humans of indeterminate sex wearing brown paper bags over their heads and on which their names were written in texta. As the procession spilled into the showground it disrupted the beer-belly competition; the contestants, even at the last minute, were desperately trying to put on form.

There were camel rides, sideshows and an opal booth. The place was packed, including many Aborigines from outlying settlements. It seemed that everyone was wearing a broad-brimmed hat, moleskin trousers and R. M. Williams boots. And it was there, in that great melting pot, where a goodly contingent of people from nowhere mixed with Croatian miners, Greeks, Italians and indigenes, that I declared the festival open; but everyone was having far too good a time to take notice of anything I said.

I enjoyed immensely my work with the opal miners, but as far as the evolution of kangaroos went, my work in Queensland, Lightning Ridge and Coober Pedy had turned up no leads. It was in more recent sediments, preserved in other regions and dating to the last 30 million years, that answers would be found.

9

The Brightest Place on Earth

From space the dazzling white salt-crust of South Australia's Lake Frome is said to be the brightest spot on Earth. Lying just east of the Flinders Ranges, the lake occupies a sunken region of saltpans and sand dunes unknown to most Australians, for it is dwarfed in both reputation and size by its westerly neighbour, Lake Eyre. But the Frome Basin is important, for here, through a process of wetting, drying and relentlessly blowing wind, the lakes have eaten deep into rocks laid down 20 to 40 million years ago—the middle of the Cainozoic era, the age of mammals.

Fresh outcrops of fossil-bearing rocks in the Frome Basin are the brightest shade of green to be seen in that arid region, and the fossils of moisture-loving creatures entombed there testify to an inland that was once verdant. That vanished landscape was a place I very much needed to understand, for I suspected that it was in some superannuated Centralian rainforest that kangaroos took their first hop.

The opportunity to visit came in the late 1970s courtesy of Tom Rich, who was mounting an expedition to Lake Tarkarooloo—one of the smaller salt lakes in the region, and the location of a previous dig. I had begun my Masters studies in geology, while Tom was pursuing his dream of finding mammals from the age of dinosaurs. Things were going badly on that front, however, and Tom felt an obligation to his employer, the Museum of Victoria, and to his country of adoption, to provide something more than his arid wanderings in Mesozoic-aged rocks had yielded. Working on the oldest mammal-bearing sediments then known in Australia would at least provide some public benefit, Tom believed, so he planned several extended periods of field work there.

Our expedition consisted of half a dozen volunteers packed into two Land Rovers, and we drove from Melbourne via Broken Hill through a countryside that had received heavy rains. On leaving the highway and entering the maze of dirt tracks that led to the lake we found ourselves in a sort of Eden. Immense fields of daisies blanketed the sandy soil, filling each swale from horizon to horizon with bright golden flowers, while the dune crests were a great mass of purple. Meandering lines of gold and purple as far as the eye could see were punctuated only by skeletal mulgas, whose deep roots would not taste the slowly percolating water for months to come, and which served to remind us of the usual order of things out here.

It seemed a violation to drive or walk over that carpet of blooms, for you crushed delicate flowers and stems at every step. And the scent of desert dust was replaced with a sweet smell of nectar and young, green growth—a heady perfume that became overwhelming as evening approached. Instead of retiring to bed with the travel-weary on our first night, something drew me into the darkening desert. Crossing one dune after another, with the light of the campfire left far behind, the sky seemed so vast and studded with stars that my mind could liken it only to a great city at night seen from the air. The stars sparkled with such vitality that you felt you could reach out and touch them, and the dizzying, sweet night-scent of a billion daisies had reached fever pitch, a billion sexual organs frantically courting a moth to their delicate stamens before the drying soil shrivelled them, robbing them of their fertility. But moths—indeed insects of any sort—were rare on that still night, for too brief a time had elapsed following the rains for them to migrate or complete their life cycle.

At first I did not notice the moon breach the horizon at my back. But as I watched it rise in the sky I witnessed a miracle in the golden fields before me. Each gilded flowerhead was turning its face to the light of the moon, until they formed a field of shimmering gold, the twice-reflected light making it seem as if the land was illuminated from below. I could not think of sleep, so I lay down among the fleshy young daisies, embraced by their scent and the still-warm air, to watch the watchers as they tracked the great silver orb on her steady journey to the far horizon.

You can live a lifetime in the Centre and never see a really big wet. The one in 1974 had filled Lake Eyre by dumping enough rain to nourish a rainforest, an event never before witnessed in living memory. But wets, both great and small, are just bounteous punctuation marks of uncertain frequency and duration in a country that nine years out of ten can stifle all but the most superbly adapted desert specialists. On that trip all I saw of red kangaroos were piles of bleached foot-bones discarded by roo shooters at gates, for the great reds that flourish in that area had dispersed far and wide to feast on the green pick.

A big male red kangaroo can exceed ninety kilograms, and there is barely an ounce of fat on his frame. As red as the sands of the Centre (the female is a smoky blue-grey—almost the colour of galahs' wings), he is the creature most people conjure up with the word 'kangaroo'. He may stand taller than you, and look down his long Roman nose at you with large brown eyes, through lashes whose length and beauty would make a diva envious. The musculature of his upper body, visible through the fine, pale fur, is that of a wrestler, while the span of his hands with their five long, sharp claws, may far exceed your own. His lower body is the essence of elegance, his long, slender legs looking almost surreal as he props himself on toes and tail-tip. If attacked, he will defend himself—with his back to a tree if possible, by grappling you close, then kicking out with both feet while balanced on his tail.

But the real miracle of the red kangaroo is more difficult to see, for the species is built to endure the erratic rhythms of the inland, when nothing can be wasted and no opportunity missed. Perhaps because of their need to travel great distances reds are not very social animals, living in smaller groups than other plains-dwelling species. But this self-reliance pays off when somehow, in the vastness of the inland, they detect that rain has fallen. Then they will move, even from an area that they have lived in all of their life, to the invisible green feast beyond the horizon. With their no-nonsense approach to life they have also dispensed with elaborate courtship. Their preliminaries to mating are rudimentary even by kangaroo standards, consisting of a golden shower (the female urinates on the male's nose), and some sniffing and pawing of the female's tail-base by the male. She favours the largest and most powerful mate she can find, and the pair copulate just once (as opposed to the repeated efforts of other kangaroos) for a brief ten to fifteen minutes.

As if they have stripped down every aspect of their existence in pursuit of flexibility and rapid growth, this business-like approach to life continues after conception. The very day she gives birth, the female red kangaroo will eject her older joey (if she has one) permanently from the pouch, and copulate so she can conceive again. Except in the most severe drought the female is never without a pouch-young. Most of her babies, however, never reach maturity, for unless conditions are favourable there is insufficient milk to feed the growing joey after it is more than a few months old. If no rain falls it is sacrificed and the next joey takes its place. The constant replacement of her growing young by the mother red kangaroo may seem brutal, but consider the sheer efficiency of this system compared with our own. Human conception is often the business of many months, while gestation takes nine months and lactation (in traditional societies) three to four years. Droughts would have come and gone, and come again, before our species had completed one reproductive cycle.

If the young red kangaroo gets past this critical juncture it grows rapidly, making its first forays from the safety of the pouch 100 days in advance of its grey-kangaroo cousins, and is weaned at about twelve months rather than eighteen months as with the grey. Having reached independence, however, the rush is over. Reds reach sexual maturity at the same age as other kangaroos, and may live longer. The oldest kangaroo on record is a male red who was tagged when fully grown in western New South Wales, and shot twenty-seven years later and over 300 kilometres away in South Australia. In effect the life of the red kangaroo mimics the rhythms of the inland; they rush to independence in the brief good season, but then live long enough to see another great wet that will carry their line onwards.

The adult red displays other abilities that have ensured its survival. An adult male is not bothered by a single dingo, for he can outrun a dog or even a horse, and if caught is likely to beat the wild dog in a fight.

Nor does competition for food in a drought usually worry them, for they can go without drinking far longer than any domestic stock, allowing them to feed over a wider area. Their food requirements, furthermore, are not overly demanding: a red kangaroo thrives on two-thirds the food required by a similar-sized sheep, and can get by with even less in a dry summer. At such times reds can eat saltbush without becoming too thirsty, and they can avoid the need to urinate by recycling urea through their saliva and into their stomachs, where it is converted into food. And then, if distant rain falls, they will all move off, leaving grey kangaroos and sheep alike to perish around shrinking and depleted waterholes, to return to the field of bones when the drought has broken.

An absence of older fossils indicates that the red kangaroo sprang into existence in the past million years. There is no doubt that it is closely related to the antilopine kangaroo (*Macropus antilopinus*) of Australia's tropical north, and may have developed from some isolated population of antilopines living on the desert margins. An Australia without reds is unimaginable, but weather shifts predicted for the continent as a result of global climate change offer a sobering reminder of how delicate the balance of life is. Many climatologists calculate that over this century the number of very hot days will increase dramatically, as will the length and severity of droughts. At the same time, rainfall over much of southern Australia will decrease—perhaps by as much as 40 per cent. Red kangaroos are creatures born of the climatic oscillations that characterise Australia, but extreme swings in climate could have disastrous consequences, for reds rely on the odd good year to replenish their numbers, and if these become too widely spaced the population will crash.

The year I first visited the Frome Basin, 1978, was a good one for reds, with few young perishing that season. Yet the ground around the lake was still parched and barren. We chose to camp beside a clump of tortured-looking coolibahs, the only shade in the area. The area that Tom intended to work ('Tom O's Quarry') was around ten kilometres from the camp, so each morning we would make lunch and drive to the site. The place was a couple of knee-deep holes a few metres across, located beside a dry salt lake, itself set in a horizontal landscape of salt scalds, samphire and grey-green clay. We would sit there scratching cautiously at the sides of the excavation, hoping to uncover some treasure—maybe the jawbone of some long-extinct koala. But such finds are as rare as gold nuggets, and a succession of the more common fossil fishbones, crocodile teeth and turtle shell fragments provided the only excitement for days on end.

By 10 am the sun had heated the air in the pits to uncomfortable levels, and at 11 we would rise to brush the dust, salt and mud from our skin and break for a cup of tea. It was then that we would confront the full force of those undersized devils of the outback, *Musca vetustissima* (in Latin 'most hairy fly'), aka the bush fly. It is much like the house fly but smaller, more persistent and far fonder of orifices—each time any of us opened our mouths to speak, the son of a maggot born on a dungheap would dart inside. Because we were host to hundreds if not thousands of these irritants, we seemed destined to swallow a fair proportion of the horde each day. Although it is possible that some entomologist actually counted the hairs on the bush fly to ensure the appropriateness of its Latin name, it seems far more likely that instead they merely estimated its hirsuteness from the torturous tickling that these licorice-tinctured insects inflict on the oesophagus.

Tom was well prepared for this particular torment, for he had recently returned from Saudi Arabia, where he had acquired the full Bedouin outfit of *thobe* and *gutra*. He was much taken with the Arabs and their way of life (though references to Noah and the flood made his search for dinosaur bones a touchy subject) and he saw nothing wrong with donning the white *thobe*, which stretched from neck to ankle, nor wrapping his head in the red-and-white checked fabric known as the *gutra*. Thus swaddled, Tom sat unperturbed, immune to heat and flies alike as he scratched for enlightenment in the quarries of Tarkarooloo, with only his American accent betraying him as a newcomer to Arabia Deserta.

After a week of unproductive torment the clouds grew dark and threatening and the wind began to gust. Tom decided that it was time for us pansies to return to camp. He had, however, just discovered two splendid fossilised turtle carapaces which he intended to excavate and encase in a plaster jacket before the storm struck. He asked that one of us return in two hours, conditions permitting, to pick him and his turtles up. If the rain bucketed down, though, we should not bother—the walk back to camp was a pleasant stroll for one encased in *thobe* and *gutra*.

When he turned up later that day, Tom told us that after we upped stakes he heard a wheezing old truck bump along the track to the quarry—the first visitor at the remote location. As the battered vehicle drew near it slowed to a halt and its driver, a rabbit shooter who sat propped between his rifle and a slab of Victoria Bitter, stared at Tom, who had entirely forgotten the strangeness of his habiliments, and asked in a tremulous voice, 'What are you doin' out here, mate?' The thunder pealed and the first fat drops of rain began to fall as the swaddled figure, without car, horse or even camel to support him, replied in a broad American accent, 'I'm digging for turtles.' At this the rabbiter gulped, turned mechanically away, and drove on up the track.

Back at camp we were sure that we were in for a soaking. I wandered

Tom Rich, enveloped in *thobe* and *gutra*, levers basalt blocks from the fossil soil horizon at Hamilton, Victoria, in the late 1970s.

out among the dry cane-grass on a sand-dune, where between thunderclaps I heard something deep underground. It was the unmistakable croaking of frogs, yet below the surface the sand was still as dry as a mummy's handshake. That afternoon the heavens opened and we copped four centimetres of rain in a couple of hours. I poked my head out of my tent in the fading light to discover that we were now camped in an inland sea populated with thousands of frogs. Their croaking was deafening, for they were everywhere. I wanted a close look, so I placed three in a billy that I hung from a coolibah, there to await inspection by the light of day.

The raucous croaking made it almost impossible to sleep, but about midnight a distinct shift in the sound occurred. Earlier it had come from all around, but now it was distinctly louder just north of my tent. The frogs were concentrating in one spot, and by morning all croaks—except for the three amplified ones coming from my billy—emanated from the one place. Beyond the camp the water still lay in a broad sheet, so I had no idea what brought the frogs together. In a few days, however, their strategy became clear. As the water receded, one pool after another dried up until there was only one left—exactly where the frogs had migrated that first night. Just what prescience led them to the spot is still a mystery to me. The next morning I discovered that my three captives were pale, flabby, golfball-sized creatures, bright of eye and with a determined turn to their wide mouths. I later identified them as water-holding frogs (*Cyclorana platycephalus*). They are among the hardiest of all amphibians and live buried in a state of suspended animation for years in the salty inland deserts, then emerge during a wet to feed and reproduce before the diminishing puddles drain away into the sand.

The Oldest Kangaroo

That expedition was cut short by the rain, which made the country impassable. Despite this I did locate some fossils from Lake Tarkarooloo in the Museum of Victoria's collection—they had been under my nose all the time. Although they were just a few teeth and foot-bones, it was clear they once belonged to animals similar to rat-kangaroos. By Lake Tarkarooloo times kangaroos were well on the way to becoming kangaroos as we know them, so I would have to search elsewhere to find the 'missing link' of kangaroo evolution—a species between the kangaroos and their possum-like ancestors.

The chance to do just that came a few months later via a visit from a famous American palaeontologist to whom Tom wished to show the outback. Lake Pinpa, a few tens of kilometres east of Lake Tarkarooloo, was to be our destination. I had seen 'Dinosaur' Jim Jensen before, and like Sandy McTaggart, in a *National Geographic* magazine. A full-page photo was dedicated to Jim as he lay stretched out beside the limb-bone of the largest dinosaur ever discovered. Yet it was only when I took in all six feet four of Dinosaur Jim and shook his hand that I realised how big that bone really was. Tom had confided that Jim was a good Mormon, and I anticipated having to stay on my best behaviour. Jim, however, bore little resemblance to the suit-wearing brethren who beat a path to my family home. As we sat around the campfire on that first evening I commented on the scattering of mosquitoes that buzzed about, the last recruits perhaps, of the big rains. They were nothing, Jim asserted, compared with what he had endured in Alaska. There, he said, the mosquitoes were so big they 'could stand flat-footed and fuck a fowl'.

Jim's experience of fossil hunting the world over was completely awe-inspiring. He had visited every continent in search of fossils and had worked with such legendary figures as Alfred Romer and George Gaylord Simpson. He had even prepared for display Australia's most famous fossil—the skeleton of the whale-sized marine reptile *Kronosaurus*. It had been discovered in the 1930s in Queensland, and now has pride of place in Harvard's natural history museum. To protect it during transportation, the huge specimen had been packed in wool scraps—dags mostly—and when it arrived in Boston harbour, in summer, it stunk so high that it had almost been destroyed as a risk to public health. I learned a huge amount from Jim about how to find and clean fossils, but it was his gift of a first-hand history of my science that has stuck most strongly with me.

Despite the fact that recent rains had washed out gullies around the lakes, exposing fresh sediment, fossils of land animals again proved hard to find. We concentrated our search on Lake Pinpa. The prettiest lake in the whole region—a mere football field of salt in comparison with most—its name derives from the Aboriginal word for the *Callitris* pines that grow in gullies along its western shore. Such bright green trees are rare in this part of the desert. Below them, low on the shore, is a layer of sediment almost as green in hue, and it is this which attracted us, for the combination of the fossils it had yielded and its stratigraphic position suggested that those clays were the oldest mammal-bearing rocks from the age of mammals then known on mainland Australia—perhaps as much as 40 million years old.

We rose each morning to cook breakfast, then commenced crawling—and I do mean crawling—over those green clays, examining every speck of colour in the hope of turning up the bone of an early marsupial. For days we shuffled on all fours, carrying our caravan of flies as we searched for the tiniest clues of past life. As in most of the Frome Basin sediments the bones of aquatic creatures—turtles, fish and lungfish—were abundant. Even the skeleton of a long-vanished fresh-water dolphin had been found there. But the remains of land-based life were rarer—perhaps a thousand times rarer—so much crawling was required before a single such fossil was found. We discovered that, strangely, the bones of land animals were often found associated with a whitish stain. We had no idea of its origin, but scientists have since suggested that it is the last remnant of decomposed crocodile faeces, voided far from shore by the marsupial-eating reptiles.

A few years earlier Tom had discovered a small tooth at Lake Pinpa which resembled the serrated premolar of a primitive kangaroo. Several kinds of marsupials, both living and extinct, possess such teeth, so we needed additional fossils—preferably bones from the hind-limb or an entire jaw of this elusive (and still unnamed) creature, to identify it— and if from a kangaroo, decide whether at this early stage in their evolution the ancestors of the living kangaroos hopped or not.

Hopping entails considerable risk of mishap, for it creates such enormous stresses that hind-limbs have to be modified in extreme ways to minimise breaks and dislocations. The pelvis and ankle are particularly vulnerable, for this is where the stresses become most extreme. In the marsupial kangaroos, the two sides of the pelvis are strongly welded together, a solution not readily adopted by placental mammals, which must give birth to large infants and so need flexibility in the pelvis. This may explain, in part, why hopping has not evolved among the larger placentals. And, as we have seen, the kangaroo ankle is highly modified to allow movement in only one direction—the slightest sideways slip and you'd have a crippled kangaroo.

Searching for such rare fossil bones in the Australian desert may seem a hopeless task. Maybe it is, and perhaps I was just extraordinarily lucky, but after several days of crawling around on the salty margin of Lake Pinpa my eyes lit upon a tiny square of bone. No more than three millimetres on each side, it was about the size of a match-head. As was my habit upon coming across such minute treasures, I sucked the dirt and salt from it, for the tongue is a far more delicate manipulator than the hand could ever be. When I carefully removed it from my mouth, glistening and clean (and thus moistened, less likely to roll away), I could see tiny square facets on it, including the distinctive stepped facet that helps make the kangaroo ankle so stable, and which is unique to them.

Here was the ankle-bone of a kangaroo unlike any I had ever seen, and I had studied quite a few. What made it special, apart from its tiny size, was the joint in front of the stepped one—the one where the bones of the foot articulate with the ankle. Instead of being oriented straight across the foot (and thus creating a stable platform), it sloped at an angle, much like the facet does in a possum. Had I found the missing link between kangaroos and possums? This we cannot tell without more evidence, but I *was* certain of one thing: the rat-sized creature that owned it may have been able to bound, but its sloped foot-joint could never bear the stress of hopping.

It is a strange thing to crawl for days on end along the edge of a salt lake in one of the most empty deserts in the world, your eyes strained with the glare of the sun and the effort of concentration, your back a vast, slow-moving vessel crewed by a thousand flies. But it is even stranger to pluck a speck of ancient bone from a decomposed crocodile turd, apply it to your tongue, peer at it, then rise to your feet screaming in exhilaration. Yet that is what happened the day I found the bone of the grandfather of all kangaroos.

When trying to imagine what kind of creature possessed that ankle-bone, we can be guided by a tiny, still-living inhabitant of the north Queensland rainforests. The musky rat-kangaroo is more rat in appearance than kangaroo, and at only half a kilo it is just twice the size of a house rat. It scuttles about the rainforest floor in broad daylight in search of insects and fallen fruit, and is the only kangaroo to do so. But to examine its peculiar anatomy you need to have one in hand. Close up, its tail is about as thick and long as a pencil and covered in coarse scales. The tail is held straight out as it scampers about and plays no role in locomotion. On its foot you will see a grasping first toe, similar to that of possums. It helps the musky rat-kangaroo scramble over fallen logs and up inclining tree-trunks. But no matter how long you observe one, you will never see the musky rat-kangaroo hop, for it is the only living kangaroo incapable of doing so, and must instead get about on all fours by bunny-hopping.

Despite the gratifying insights that the musky rat-kangaroo can bring to the study of ancient bones, that earliest kangaroo fossil remains as much a frustration as a source of enlightenment, for to this day we have no idea how long ago it lived. Over the past quarter-century there has been plenty of arm-waving, guessing and tenuous reasoning to support a view that it might be around 25 million years old, but it could be

40 million, or as little as 20 million years old. Our incapacity to date the fossil-bearing sediments directly holds true for all but one of the mammal-bearing fossil deposits found on mainland Australia older than five million years, and until we have a way of answering the 'when' of palaeontology we will be unable to adequately address the more interesting 'why' and 'how'.

We can, nevertheless, say a little about the vanished Australia that was home to this miniature kangaroo. We know from other fossils discovered at Lake Pinpa that at the time the roo lived the Frome Basin was a vast, freshwater lake which must have had some connection with the sea, for long-beaked freshwater dolphins descended of seagoing ancestors abounded in its waters. Fish, crocodiles and tortoises also shared the lake with an astonishing nine species of lungfish, the largest of which reached three metres in length. There were ancient, toothed platypuses, grebe-like birds, pelicans, shags and ducks, all species now long extinct, while curlews, songbirds and gulls populated its shores. A touch of the exotic was added by flamingoes. And there were tortoises, resembling those of the Galapagos but with armoured tails and cow-like horns on their head, grazing beside that ancient lake.

The marsupials were, by and large, unlike anything living today. The largest were sheep- to tapir-sized creatures known as wynyardiids and ilariids. The first takes its name from the sleepy town of Wynyard on Tasmania's north coast, whose scientific claim to fame is a rocky outcrop near the town's golf course where in the late nineteenth century the skeleton of a brushtail-possum-sized animal was found. Named *Wynyardia bassiana*, the skull and jaw had been so exposed to the waves of Bass Strait that every tooth had been eroded away, but the bones were found in sediments that yield tiny marine organisms, allowing us to date this particular skeleton to around 23 million years old.

Frustratingly, however, not one more scrap of land mammal bone has been found at the outcrop since.

The ilariids' name is from an Aboriginal word meaning 'strange', and a highly appropriate name it is, for with no close living relatives these primitive herbivores (represented by several species) may have resembled ponderous terrestrial koalas. Odd-looking *Miralina* and *Ektopodon* possums were also present. Around the size of a brushtail possum, *Ektopodon* was as short-faced as a cat and probably a fruit-eater.

Ringtail possums, a strange pygmy possum, an ancient koala and our miniature, ancestral kangaroo complete the rollcall of the marsupials of that vanished era. We know almost nothing of the vegetation that fed these creatures, though we can surmise that the Flinders Ranges supported hanging bowers of verdant rainforest. On the plains around the lake a mixture of herb-filled glades and gallery rainforest may have grown. To admit the slenderness of our knowledge of this vital period in Australia's prehistory is painful for one who has spent a lifetime studying fossils, but I must concede that it will fall to future generations to see the details of this age with true clarity.

Skeletons in the Dead Centre

A region of sand dunes and expansive salt lakes lies at the heart of Australia. Lake Eyre, the largest of them all, extends 200 kilometres from north to south and sixty-five east to west—the size of a small European country. While the lake itself has yielded few fossil deposits, smaller nearby lakes hide a fantastic wealth of materials, a bounty which has drawn me back repeatedly since my first visit in 1980.

The region is best approached by driving north from Adelaide on a road that passes west of the Flinders Ranges, their sublime peaks rising like ramparts out of the desert plain. In the middle of this grand land-scape lies Parachilna, an old stop on the now disused rail-line to Marree. The settlement has recently found a second life as a stopover for tourists who can enjoy a camel-and-emu sausage or kangaroo steak before retiring to the comfort of clean sheets. As we pulled into Parachilna on a visit in 1997 a sign caught my eye. Written with texta on a piece of cardboard propped beside a humpy, it read 'Kev's Camel Capers'. A few camels hobbled in the distance and a man, the presumed master of the 'capers', lay prone in the dust under a low bough shelter. One of our team introduced himself and asked how business was going. 'Better than the dole' was Kev's reply. He then explained that he got his animals from a camel butcher who caught them up around Lake Eyre. Asked if he had any plans, Kev responded, 'Yeah. Got a Filipino bride coming out next month. Got her in Manila. Advertised that I ran a transport business.' And, from what I heard at Parachilna, she would not be the only Filipina to have moved to the outback.

After a camel sausage we headed to Marree, the only town in the salt and gibber country east of Lake Eyre. I was hoping to meet Old Achmed, the last of the Afghan cameleers, those hardy frontiersmen who provided transport for explorers such as Warburton and Giles, and who for almost a century supplied remote settlements with goods and news of the outside world.

Marree is little more than a hotel, a store, and a motley collection of buildings clustered around the disused railhead. I found Achmed in the 'coffee shop' (which was also the general store, petrol station and anything else it needed to be). He was full-faced, had a rounded figure and seemed to be a youthful eighty-something—hardly what I had expected.

Achmed told me that until 1956 he had worked with his father carting beer on camel-back from the railhead at Marree to the Birdsville pub—a distance of about 450 kilometres—and being good Muslims they were probably the only people in the outback you could trust with the precious cargo. It was an incredibly tough life, the two-week journey crossing a region along the edge of the Simpson Desert treacherous to negotiate. Then, in 1956, a truck started to make the run and put the cameleers out of business. Achmed was forty-two years old, and for the first time in his life found himself out of work.

On hearing his story I made sympathetic noises, imagining the difficulties he faced in retraining, but Achmed would have none of it. 'No, mate,' he expostulated, 'it was the happiest day of me life.'

'But how did you make up the income?' I asked.

'Oh, Dad never paid me, so there wasn't much difference, except for the work.'

The community at Marree is mixed—Aboriginal, Afghan and European—and from the pub I saw a couple of Aboriginal youths, golf clubs slung over their shoulders, heading for a graveyard of old cars. When I asked the publican what they were doing he looked at me as if I were an idiot and spelled out slowly, 'They're going to play a round of golf, mate.'

'But where's the course?' I replied, looking out on the bulldust and rusting metal.

'That's it there. Can't you see the flags?'

Sure enough, rising from among the rusting chassis were flags of the kind that mark tees. Marree is a strange place.

In 1980 I had no idea what to expect from Lake Eyre. Despite being known as the rain gauge of Australia (one sixth of the continent drains into it), the Lake Eyre Basin is the driest part of the country. But when the tropical monsoon brings rain to the headwaters of the Cooper Creek in far away Queensland, the stretch of salt crust surges into life, becoming an inland sea whose fresh waters float atop the brine. Then, birds and fish in their millions thrive in the parched interior, which may itself have received no rain at all.

The region has attracted the attention of palaeontologists ever since Scots professor J. W. Gregory led a group of students from the University of Melbourne there in the summer of 1901. University holidays seem perversely scheduled to torment those who are desert-

Encasing fossils in plaster for safe transportation is a filthy job. As Tom's volunteer I crawled under this block to plaster the overhead base. I'm wrapped in hessian to keep the wet plaster, dust and flies off my skin.

or tropics-bent, but Gregory's team did remarkably well, finding an extensive outcrop of fossil bones on Cooper Creek. But these were recent specimens—none more than a million or so years old, and I was interested in far older fare. (Gregory, incidentally, drowned in the Amazon River, an odd end for a man whose reputation was made in deserts.)

On that trip in 1980 I camped with Tom Rich and his wife Pat, also a palaeontologist. We had come to look for fossils at Lake Palankarinna, but had little luck. To the east of our camp was gibber—the highly polished pavement of desert stone so characteristic of large parts of the Centre. Traversing it one morning we came across a pair of dingoes that had pulled down a female red kangaroo. The blood was still wet on the stones when we stopped to investigate, and the clay between them told of the doe's dawn struggle for her life. Dingo numbers are so high in this cattle country that reds are a rarity.

Within a few days of our arrival black cloud had built up, sealing in the heat and producing an unpleasant humidity. Upon stepping out of my tent around midnight for relief from the sauna-like atmosphere, I found the ground alive with scorpions, the largest of which were twenty centimetres long. The place was seething with them, with barely room enough to place a foot between the pale, translucent bodies. Perhaps they had emerged to mate or hunt; whatever the cause I have never seen a sight like it before or since. In the small hours a thin scatter of raindrops fell, yet by dawn all signs of the ghostly-pale scorpions and moisture alike had vanished.

In 1982 I had the opportunity to work at the most remote fossil locality in the Simpson Desert. Dating to a more recent geological epoch than those from Lake Pinpa, it outcrops around Lake Ngapakaldi, a small salt lake lying to the northeast of Lake Eyre. Getting there is difficult, involving grinding across the bleached sand of the Cooper Creek, then picking the inter-dune valley that the lake lies in. Choose the wrong valley and you could drive right past without ever seeing the small oval of salt.

The whole region north of the Cooper was as dry as a sunstruck bone during our visit and, with the exception of a teeming mass of rabbits in this pre-calici Australia, appeared lifeless. As we set up camp just south of the lake, rabbits swarmed around us, making me determined to enjoy a roast dinner before we quit the place—a pleasure we arranged that very evening. Dawn comes strangely in this part of the desert. The gradually lightening sky leads first to an awareness of the dry air, and as I lay choking on a dry throat not a single bird gave voice to greet the day. The deathly silence was broken only by a distant pulse—the sound of my own blood pumping.

As we walked the two kilometres that separated us from the lake we saw no spinifex, but instead a miserable sort of cane grass that favoured the dune summit, while a few dead-looking acacia and rattle-pod bushes lay scattered over the inter-dune valley. We continued in silence, carrying tools and water, and I came to appreciate how this most remote of ecosystems had been devastated by European impacts, despite the fact that few Europeans had ever entered it. Too dry to be used for cattle, it had become a paradise for rabbits, and doubtless the foxes and cats that prey upon them. They had eaten the place alive, and all that was left of the native mammals that had thrived fifty years earlier was the odd bone poking out of the dunes. Even the birds and insects had been affected, for the rabbits had stripped much of the vegetation. I have often wondered how this corner of Australia looks now, after calicivirus has purged the place of rabbits, and whether any remnant of those vanished marsupials might have survived to enjoy the arid paradise that, I hope, has sprung up in its wake.

Lake Ngapakaldi's only distinguishing feature is a scatter of water-worn pebbles on its shore, an unusual thing in this desert of wind-eroded stones, and it marks the bed of a creek that last ran perhaps 20 million years ago. If you dig down deep into the briny, pebbly sand you will hit a layer of thick black clay. Break open a lump, crumble it under your nose, and you will smell something utterly different from the dusty, salty and bitter smell of the desert; rich organic compost from the floor of a central Australian forest which flourished 20 million years ago.

Whole leaves have been held in suspended animation by that remarkable clay, and although they have turned black they remain as flexible as the day they fell from the twig. But when the leaves are again exposed to sunlight an eon of decomposition catches up with them, and in the time it takes to boil a billy they are nothing but dust. Scientists have beaten the heat and dryness by preserving a few fragments in glycerine, which reveals that Lake Ngapakaldi's ancient billabong was fringed with reeds, behind which grew trees typical of the rainforest edge; among them was a genus that was to achieve dominance—the eucalypt. Their presence is a clear indication that by Ngapakaldi times the rainforests were breaking up and the seeds of a future Australia were springing up at the continental heart.

The fossil bones we had come to find were concentrated in gravels below the clay. Most were small and water-worn, as if they had travelled a great distance in that ancient stream before coming to rest. Sometimes I would spot the distinctive brown colour of bone protruding from a clod of gravel and break it open to reveal the jawbone of an ancient ringtail possum. The work was like time travel and I was entranced, sweating as I swung the mattock. Another clod might reveal the jaw of a bandicoot, or a wallaby-like creature, the likes of which the world has not seen since the Miocene period 20 million years ago, and as I worked on, in my mind I was soon back in that very different central Australia.

We did not see most of the fossils we excavated, for the gravel had to be dried, broken down in water and picked through under a magnifying glass before they would be revealed. So we bagged up the sediment and loaded it into the Land Rover to await a cool night, preferably with a dew to consolidate the sand. Over the days our stockpile beside the Birdsville Track grew and grew, until a truck picked it up and delivered it to the railhead at Marree, and thence to Sydney.

Me as an eighteen-year-old in the middle of a sweltering summer, unearthing diprotodon bones for Tom Rich at a location near Bacchus Marsh, Victoria.

The western grey (left) and eastern grey kangaroos (top); Bill's and my Moto Guzzis on the Nullarbor Plain, 1975 (above). Note the esky strapped to the back of my bike.

Clockwise from top left: quokka, bridled nailtail wallaby, musky rat-kangaroo, euro.

Clockwise from left: bugaree (swamp wallaby), burrowing bettong, red kangaroo–the symbol of Australia.

With dingiso, a black-and-white tree-kangaroo,
Tembagapura, West Papua, 1995.

Where the Great Roos Came From

In Sydney we sorted the sack's contents at Michael Archer's laboratories in the University of New South Wales, and found that the fossils from Lake Ngapakaldi were typical of those from central and northern Australia which date to the Miocene Period, 24 to 5 million years ago. In 1982 little was known of the kangaroos of this period—just a few teeth, jaws and isolated bones—and they presented a great puzzle. While many teeth looked superficially similar to those of later kangaroos, subtleties of their shape—a cleft in the enamel crown here, a wriggle there—intimated that the same tooth shape may have arisen in entirely different ways in these ancient kangaroos as compared with living species. Yet somewhere here lay the origins of the great kangaroo subfamilies—the Macropodinae and Sthenurinae—that would dominate ice-age and modern Australia.

The ancient kangaroos from Lake Ngapakaldi and similar-aged sites could be divided into two types, which myself, Michael Archer (then my PhD supervisor) and Mike Plane of the Bureau of Mineral Resources classified into two subfamilies that we called the Bulungamayinae and Balbarinae. The balbarines were generally larger (about the size of hares), and had molars resembling those of the quokka. We decided that the balbarines were the most likely candidates to be the mothers of all kangaroos, the rat-kangaroos alone deriving from the bulungamayines. But we were soon to learn that the evolution of the kangaroo family resembles a great twiggy bush, where similarities can sometimes be misleading.

The discovery of rich fossil deposits at Riversleigh finally dispelled all doubt surrounding this aspect of kangaroo evolution. Michael Archer had a long-standing interest in Queensland, for he had been curator of mammals at the Queensland Museum before coming to Sydney and, in 1981, he invited me to visit some fossil deposits on Riversleigh station in Queensland's Gulf country.

Our route lay through Mount Isa and during our overnight stay there we witnessed several brawls; but the sense of intimidation they produced was nothing compared with the feelings the countryside elicited as we travelled towards Riversleigh station. The tropical heat and blistered rocks give the impression that, if lost in that country, there would be no surviving even a few hours without water. Then, almost miraculously, we came across a hidden oasis—the limpid, spring-fed Gregory River. Our base camp was established among the deep green trees of this paradise where waterlilies, freshwater crocs and waterbirds formed an ever-changing panorama. But the place we had come to investigate was not in this verdant strip but on a blasted limestone ridge a few kilometres away.

The following morning, on passing the Riversleigh homestead we discovered that 'Honest John' Moloney, a travelling vendor to stockmen, had set up shop at the front gate. It was rodeo time and the young jackaroos had gathered around Honest John's van, for it was important to look the part when atop a wild bull or bucking bronco, and these men were paying large dollops of hard-earned cash for R. M. Williams belts, boots and hats. The whole ritual, from Honest John's spruiking of his goods from the back of his gaudily painted van, to the shy Aboriginal youths awaiting their turn to buy, was a vivid reminder that we were in cattle country.

We continued on, finally reaching a low limestone cliff to see where fossils had first been found in 1910. Impressive remains still abounded, in one huge boulder the jaws of massive crocodiles jostled for space with the skeleton of a titanic, emu-like bird known as a dromornithid. Yet the limestone was tough and mammal fossils scarce so we set to work at an old bat roost discovered four years earlier. We broke it apart with explosives, then gathered the fragments and carried them back to the University of New South Wales to be dissolved in acetic acid, thereby releasing the bones.

This Riversleigh expedition was a disappointment to me. At that time only a couple of kangaroo jawbones had been found in the deposits, none of which contributed much to our understanding of kangaroo evolution. But what was worse from my youthful perspective was that there was so little to see at the site, for the major fossil deposits were yet to be revealed.

In 1983 Michael Archer returned to Riversleigh—and hit the jackpot. The new sites yielded thousands of kangaroo fossils, including some near-complete skeletons. Dr Bernie Cook, a palaeontologist from Queensland, elegantly analysed this material, demonstrating conclusively that the balbarine kangaroos, whose teeth look so similar to those of some living wallabies, were in fact an evolutionary dead-end. Rather than being the mothers of all kangaroos, they are a primitive lineage that branched off early in the evolution of the group and led nowhere. It was the rat-kangaroo-like bulungamayines that had given rise to both rat-kangaroos and the more successful macropodines that so dominate Australian environments today. I was not really surprised to see our earlier hypothesis overturned, for it was based on very few fossils, though all that was available.

Bernie's discoveries provided a vital lesson in the nature of the enterprise I had embarked upon. Sooner or later all scientific hypotheses will be proven wrong or insufficient. That is the way science progresses, and scientists are well served by never allowing too much ego to get tied up in their discoveries, for it is inevitable that other researchers will come along to test, modify and often dismiss their ideas.

Dr Ben Kear, who had been my doctoral student, also studied the Riversleigh kangaroos. His research, which has focused on the few partial kangaroo skeletons found, has provided insights into why the bulungamayine kangaroos were so successful while the balbarids vanished. A balbarid kangaroo skeleton lacks several specialisations necessary for hopping. Instead it retained a grasping great toe on its hind-foot (like a possum) and a partially prehensile tail. This suggests that balbarids bunny-hopped, and possibly climbed, through the rainforest. The bulungamayines may have been the very first kangaroos to hop, thus translating the potential of their reorganised hind-limb into the most efficient means of getting about ever devised by a land mammal. This alone may have assured their success, but it seems likely that, in the lineage leading to the kangaroos, wallabies and rat-kangaroos, it was accompanied by a second change which involved their digestion.

Today the macropodine kangaroos, the group that dominates the continent, share Australia with the rat-kangaroos—animals such as bettongs and potoroos. At the time of European settlement ten species of rat-kangaroo lived in Australia, and they were sometimes considered pests. Very few Australians see rat-kangaroos in the wild today, for their numbers are much diminished. This is a tragedy for the Australian environment as these seemingly obscure animals play a vital role in maintaining the health of our ecosystems. They do this by spreading a vital partner of many Australian plant species—the mycorrhizal fungi, otherwise known as 'marsupial truffles'.

The mycorrhizae are normally invisible thread-like organisms that remain buried underground. They occur in massive volume in Australian soils, and are strongly attracted to the roots of many plants, around which they form a sheath. Australia's soils are so infertile that many plants cannot support their growth unless they form a partnership with these fungi, which have a special way of concentrating soil nutrients. In the kwongan heathlands of southwestern Australia around 80 per cent of all plants have an association with the fungi. Even the mighty eucalypts depend heavily upon the thirty-micron-thick halo of fungus that surrounds their roots, allowing them to grow faster, be healthier and recover more rapidly from wilting.

For their part the mycorrhizae gain water and sugars from the plants. But they have a problem. Many are host-specific, and getting from one host plant to another is difficult. This is where the rat-kangaroos come in, for they have as exquisite a nose as any truffle-hunting pig, and can smell the 'marsupial truffles' produced by the fungi deep underground. These they dig up and consume, in the process spreading the fungi's spores. Rat-kangaroos have a greatly expanded stomach, dominated by one very large pouch at its front end, which is used like a sack to store the 'truffles'. For such small, vulnerable creatures, being able to quickly fill their stomach-sacks, then retreat to shelter, is advantageous, for it minimises exposure to predators.

Rat-kangaroos also assist the environment by turning over the leaf litter, thereby hastening its decomposition and enhancing the water-holding capacity of the soil. The boodie or burrowing bettong (*Bettongia lesueur*), digs far deeper, and in so doing unearths fertile soil from deep underground. Two hundred years ago these grey, blunt-faced, rabbit-sized animals could be found over most of the drier parts of the continent. They did not decline immediately upon European colonisation—so abundant were they around some settlements in the mid-nineteenth century that their penchant for excavating in country cemeteries saw them accused of 'disturbing the rest' of the pioneers—but by 1940 they had vanished entirely from the Australian mainland and today survive on just a few offshore islands.

Burrowing bettongs are feisty creatures whose growls and grunts, if disturbed, give adequate warning to human and goanna alike to leave them alone or risk a severe biting and scratching. Their pugnacious nature, however, is coupled with a stunning naivety in the face of obvious danger. Scientists first stumbled across this aspect of their personalities when they had to instil fear of dogs into some island-bred bettongs to be released in a mainland national park.

In order to convince all doubters, the researchers produced a film of the experiment, which shows a fearsome bull-mastiff frothing at the mouth and barking as it strains at the end of its chain, while centimetres from its nose a trusting bettong is stretching towards the open maw, wearing a contented, curious expression. It took many weeks to teach the bettongs that dogs could be dangerous, but even that hard-won knowledge was short-lived, for when the bettongs were retested a few months later they again displayed no fear of the chained dog!

The warrens of burrowing bettongs can still be seen in many parts of Australia. Areas with a hard, almost concrete-like surface layer known as calcrete were especially favoured (which may explain their predilection for burrowing under gravestones). In 2000, on Mount Gibson station, several hundred kilometres north of Perth, I saw some burrows in mulga country with very little shrub or herb understorey. But a very different plant community grew where the bettongs had heaped up great piles of earth from under the calcrete. These plants presumably need the extra nutrients offered by the fresh soil, and possibly support a greater diversity of animals than might otherwise exist. You might think that those inveterate burrowers, the rabbits, would make a good replacement for the bettongs, but their proclivity to eat the plants in the vicinity of their burrows means that the benefits of their excavations extend to few creatures but themselves.

The Riversleigh deposits have yielded the skull and jaws of an animal strikingly similar to the living burrowing bettong, suggesting that these creatures have been enhancing Australia's biodiversity for millions of years. Other kinds of rat-kangaroos, though, are much more prevalent at Riversleigh, foremost of which, I discovered to my delight, was a relative of the *Propleopus* that had so intrigued me in the Museum of Victoria years before.

The creature was the size of a Tasmanian devil (or perhaps a little larger) and was the largest kangaroo then known from Riversleigh. To my dismay the most complete jaw recovered had lost its premolar to blasting powder. Thankfully a natural mould was preserved in the surrounding rock and I made a cast of the missing tooth with resin before dissolving the stone in acid. Michael Archer and I named the new species *Ekaltadeta ima* meaning 'powerful tooth' and 'condemned to die' in an Aboriginal language of the Northern Territory, a name inspired by the missing premolar, which was clearly of massive proportions. *Ekaltadeta* is the earliest known member of the *Propleopus* lineage, and its description was a start to the studies I had so long hoped to complete, but until I could examine all of the *Propleopus* material held in the world's museum collections I would not be able to fathom the significance of those huge, buzz-saw-shaped teeth.

Mike Archer continued to explore the Riversleigh region in 1981 while I left with Tom Rich for the Northern Territory to investigate another fossil locality. Bullock Creek is on a property several hours' drive south of Katherine, and when we arrived I puzzled over the long, sinuous flat-topped hill that wound its way over the landscape. Only on close examination did I realise it was all that remained of a mighty river channel whose water carried so much carbonate that limestone had formed in its bed. Over the millions of years since the river had dried up, erosion had removed much of the softer, surrounding rock, leaving the limestone-filled channel standing at least ten metres above the plain.

In the bends of the ancient river system the bones of thousands of huge extinct animals could be seen bristling from the rock. They had been deposited at 'point bars', which are well known to those who swim in or camp by rivers because sandy spits often form in such places. This is where the current slows as the river bends, allowing things carried in the flow to drop to the bottom. The limestone enclosing the bones was white, while the bones themselves were as black as ebony, making the Bullock Creek fossils some of the most spectacular I had ever seen. Slightly acidic rain had dissolved the stone faster than the fossils, leaving the bones standing out like a bizarre Egyptian bas relief. As evidenced by their fossilised remains, huge crocodiles and giant goannas must have flourished in the region; but the most common creature by far in the deposits was a species of *Neohelos*, a relative of the diprotodon and about as large as a calf. Around 10 to 20 million years ago entire herds of these wombat-like creatures must have drowned in ancient floods, for their skeletons formed great mats in the limestone. The remains of kangaroos were, I discovered to my disappointment, much rarer. Only three rather uninformative fossils were found, all balbarines of a species about the size of a rock wallaby.

There is one final fossil locality that adds to our understanding of the changing situation in central Australia, and to the evolution of kangaroos. It is on Alcoota station near Alice Springs, and although we cannot date the two fossil-bearing layers found there, they seem to hail from a time when major deposition at Riversleigh had ceased. Peter Murray, the scientist who has excavated the site, estimates that the two layers may be around 8 and 6 million years old. The fossils preserved therein reveal that by this time rainforest had given way to more open environments—perhaps shrubfields, woodlands or savanna—that supported herds of ponderous diprotodontids and relatives, of which there are four kinds, the largest the size of a cow. These creatures shared the landscape with several species of dromornithids, including *Bullockornis*, the largest bird that ever lived. Pat Rich, who described this monster, has a cardboard cutout of its leg mounted on the wall of her office at Monash University. The top of its femur stands well above head-height.

It is also in these deposits, laid down towards the end of the Miocene, that we find evidence of the first kangaroos to reach anything larger than wallaby-size. *Hadronomas*, one of the earliest members of the short-faced kangaroo subfamily Sthenurinae, was as big as a grey kangaroo. It shared its habitat with a much smaller (fox-sized) species known as *Dorcopsoides*, which is the earliest of the macropodines, and it is with this creature that we see the piccaninny dawn of our modern age.

The Age of Kangaroos

Following the fleeting appearance of *Dorcopsoides* on the central Australian stage around eight million years ago, a frustrating dark age draws a curtain over our window on the past. When the curtain lifts, around 4.5 million years ago, a very different set of animals has populated Australia's stage. From the dainty striped wallaby to the great grey kangaroo, the macropodine genera as we know them today are all arrayed before the palaeontologist, and forever after kangaroos dominate Australia. From browsers to grazers and fungivores to carnivores, for over four million years their increasing dominance was such that by the ice age (when the megafauna roamed the land, between about 2.4 million and 50,000 years ago) Australia was home to over eighty species of kangaroo.

We still live in the age of the kangaroo. But what of its initial flowering? Aided by its truffle-loving rat-kangaroos, southeast Australia is, from its mallee to its snow gums, a land of eucalypts. Five million years ago, however, rainforests still covered enough of southern Australia to support tree-kangaroos, and most of what we know of the animals that lived in those forests comes from a fossil locality situated in the grassy woodlands of Victoria's Western District. The fossils owe their preservation to a catastrophe that occurred 4.46 million years ago when a lava flow overwhelmed an ancient rainforest, burning it off at the stumps and instantly freezing all biological activity. Very few bones are found at the site. Instead what you find are the exquisitely preserved enamel caps of teeth— the hardest part of the skeleton—all stained navy blue and as sharp as the day they cut their last food.

When I first saw the Hamilton site with Tom Rich in 1974 I was unimpressed. It lies at the bottom of a sheep paddock beside a grossly polluted creek known as the Grange Burn, whose banks have collapsed and eroded through thousands of cloven hooves. Yet this was once a pretty place, for upstream a waterfall tumbles into a peaceful amphitheatre, while a few reeds and white-flowering *Bursaria* cling to the most inaccessible niches, reminders of a more diverse vegetation. Tom recounted that the site had been discovered some thirty years earlier by the Reverend Edmund Gill, a parson of scientific persuasion and Tom's predecessor at the museum, who in his wanderings had plucked a solitary molar from the fossil soil below the basalt. The tooth, which had the appearance of a shrivelled pea, once graced the mouth of a *Propleopus*. The find had inspired a group of Americans to excavate, and they recovered the isolated teeth of many other animals, but nothing more of *Propleopus*.

Tom had brought us here because the Hamilton site was, he said, the 'Rosetta stone of Australian palaeontology'. Not only had the basalt preserved the site, but a technique known as potassium-argon dating (the ratios of these two elements give an estimate of when the basalt congealed) allows geologists to determine its precise time of formation—in this case 4.46 million years ago. There are very few precisely dated sites in Australia, and comparisons with the Hamilton site have permitted the ages of many other fossils to be estimated. Our job was to unearth more fossils so that those estimates could be refined. For my part, I was keen to learn more about *Propleopus*. The whole idea of such a creature seemed preposterous to me and, as I was to learn, in many ways the giant rat-kangaroo is one of the most bizarre creatures that ever lived. Yet so elusive was evidence of it that I had taken to calling it the 'probably hopeless' because I despaired of discovering anything more about it.

For sixteen days we moved innumerable basalt blocks weighing several tonnes apiece and sieved many cubic metres of fossil soil without finding a single fossil tooth. Tom was doing his best to hold the field crew together, buying chocolate and steak for the teenage volunteers out of his own pocket. Nevertheless, conspiratorial whispers about whether we were digging at the right spot were frequently to be heard, and there was even talk of desertion. But then, late on the seventeenth day, after most of the field crew had packed up and gone home, one of the volunteers caught the glint of ancient blue tooth enamel in the bottom of his sieve. It was not a tooth of the longed-for *Propleopus* but that of a smaller, more ordinary kind of kangaroo. Still, I considered that we had struck paydirt, and from then on was ready to toil year after year beside the polluted waters of the Grange Burn.

While working at Hamilton we stayed at the shearers' quarters on a nearby pastoral property in a rolling landscape of grassland and stately red-gums. Tom had secured this luxury in an unusual manner. Several years earlier he had been told of a grazier who was the proud owner of the skull of an Irish elk. The fossil was in need of expert attention, for its teeth were coming loose. The Irish elk was a denizen of the

European ice age and a giant of the deer world, its antlers measuring nearly three metres across. Even today a splendid set of antlers is occasionally dragged from an Irish bog, though good skulls are rare and among the prized possessions of many museums. Tom, sceptical that one might reside near Hamilton, nevertheless decided to pay a visit and sure enough, there in the hall of the nineteenth-century farmhouse hung one of the most splendid Irish elk skulls he had ever seen—much grander than the one owned by the Museum of Victoria. The grazier's ancestors, it transpired, were Irish aristocrats and had shipped the heirloom across in the 1840s. Two years after administering some dental care to the fossil, Tom raised the issue of accommodation, and use of the shearers' quarters for his field crew was readily proffered.

At first the arrangement seemed ideal, for the century-old building was comfortable and roomy, comprising a large kitchen and numerous bunk beds. But we soon learned that you had to be careful when wandering far from the door, for the place was located in the bull paddock. Everything was fine at first, with the bulls remaining in a distant corner of their domain. Then a herd of heifers was moved into an adjacent paddock. To our consternation these shameless hussies began to idle their time away beside the fence in the vicinity of the shearers' quarters, flaunting their posteriors and generally teasing the bulls mercilessly. One bull, driven to distraction by this display of bovine pulchritude, decided to leap the fence but, lacking the agility of a kangaroo, his hind-leg became entangled in the wire and was sprained. Once among the females the poor fellow could do nothing, for every time he tried to mount a willing heifer his gammy leg would give way and frustrated bellowing would fill the air. When he was returned to his own paddock with his mood considerably soured, we became even more cautious when wandering outside for firewood and other essentials.

One hot summer night we discovered that bulls were not the only hazard in the paddock. A sinister reptilian head emerged from a knothole in the wooden floor, quickly followed by a metre of tiger snake. The creature had not noticed that the room was occupied until enough of its body had passed through the knothole to make retreat impossible, so instead it decided to advance and bluff its way out. In an instant the room was awash with panicking people, which in turn panicked the snake further. Unable to locate the hole it arose from, it shot about the room full of fury and venom, striking out in all directions as it searched for an alternative escape route. Half a dozen of us shared the tabletop while one of the more nimble volunteers leapt onto the mantelpiece where he did a marvellous job of balancing for several minutes. Someone finally plucked up the courage to descend to the floor and open the door, allowing the snake to flee into the night.

One year, wanting to scout the region for other fossil localities, I arrived a few days in advance of the rest of the crew. The shearers were still in residence, and that is when I met Tom the shearers' cook, a pint-sized fellow well known in the district for his eccentric ways. When the owner's son introduced me as 'Tim Flannery, the fossil hunter', both he and I were astonished at the warmth of Tom's welcome.

'Oh mate, I've been waitin fer ages fer yer to turn up!' Tom exclaimed as if I were a long-lost brother. 'What would yer like to eat—sausage rolls?'

Soon there was a pile of them on the table—and a bottle of tomato sauce—and for a couple of days Tom and I got on famously. I took all of this to be a symptom of his eccentricity, but then—inexplicably—his enthusiasm for me began to cool, until by the end of the week he seemed hardly able to bear the sight of me. Perplexed, I asked one of the shearers what was going on. Tom, it seems, was a little hard of hearing and had mistaken my being a fossil hunter for a possum hunter. He was thus understandably irate that numbers of the furry creatures— the only marsupials in the area—had not diminished at all during my stay, and that I had kept to my cot at night when I should have been out hunting.

Tom had a special reason for hating possums. In the absence of other nesting places in that denuded landscape, they had sought refuge in the roof cavity of our quarters, where the ghastly growling of the males (which sounds like rasping, heavy breathing) was keeping everyone awake at night. This had not helped the mood in the shearing shed, and when one of Tom's stews had been impugned with suspicion of pollution from the possum-piss-stained ceiling, the cook's hatred had turned virulent. Thinking that my arrival was a last-minute bid by the cocky to avert bloodshed, he had greeted me as a saviour. But as my laziness became evident, Tom's disappointment knew no bounds. The shearers, it transpired, had caught on to the joke and were making things worse by spreading a rumour (alas, all too true) that one evening while driving back from the Branxholme hotel I had even swerved to avoid running over a possum! The Branxholme, incidentally, was a great place to drink, for the clock had no glass on its face and when the local policeman was at the bar the time always seemed to show five minutes to closing. Tom and I were never reconciled, but with the shed 'done' the shearers and their cook moved on, and Tom Rich's volunteers began arriving for another summer's work.

Advancing with Feet or Stomach?

Each field season on the Grange Burn revealed a little more of the ancient rainforests that once thrived in Victoria. We learned that the tree-stumps burned off by the basalt flow were celery-top pine, which still grows in Tasmania, yet the fossil teeth we uncovered were like nothing from our southernmost state. Upon breaking open a clod of blueish-grey soil one afternoon, I discovered the distinctive teeth of a type of wallaby known only from the mountains of New Guinea, over 3000 kilometres away to the north. And a few days after that, the tooth of a tree-kangaroo emerged from one of the sieves we used to wash the ancient soil.

Making sense of such finds took some time, but my understanding of the area grew after wandering a few hundred metres downstream to a dome of pink granite-like rock that jutted out from the creek bank. On its upstream side fossil soil lapped its base, while against its downstream face lay sediments formed in an ancient Bass Strait. Here, some eighty kilometres from the modern coast, was the ancient shoreline, complete with oysters, the bones of whales and even the teeth of great white sharks that once swam in its waters. Four and a half million years ago the pink dome of rock must have stood as a bulwark against the force of the Southern Ocean, and in its lee grew our ancient rainforest, perhaps watered by the ancestor of the Grange Burn itself.

It was now 1980, and after four seasons of digging we had turned up only eight isolated teeth belonging to *Propleopus* which, to add to my frustration, were sent to an expert for study. To me, however, fell the immense pleasure of describing the other kangaroos from the site. There were now hundreds of teeth representing over a dozen species. A close relative of Queensland's musky rat-kangaroo shared Hamilton's ancient forests with a larger rat-kangaroo of the bettong type, but the most common creature was very similar to the Tasmanian pademelon (*Thylogale billardierii*), which vanished from Victoria a century ago. Hundreds of its teeth had been unearthed, indicating that it was then as dominant in Victoria as its descendant is today in Tasmania.

The genus name for pademelons, *Thylogale*, means 'pouched weasel', which the creatures most emphatically do not resemble. They are instead rather nondescript wallabies, all six species of which inhabit the margins of rainforests and dense scrub from Tasmania to New Guinea. They may well be the ancestral type from which the great kangaroos, striped wallabies, rock-wallabies and tree-kangaroos have sprung, so in some ways they are living fossils. Studying their teeth, I was amazed to see how little the lineage had changed in 4.46 million years. So successful is the Tasmanian pademelon that permits are given to farmers to cull them. While this may seem odd, the creation of pasture by Europeans has benefited it greatly, justifying a sustainable harvest. And besides, its flesh is so far superior to that of the kangaroo that a more basic argument may persuade people to sample it. Were it more readily available, pademelon would be one of Australia's premium meats.

Almost as common at the Hamilton site were the teeth of a forest wallaby (genus *Dorcopsulus*) of a sort today restricted to the mountains of New Guinea. The remains of these hare-sized creatures have not been found at any intermediate localities, so their presence in southern Australia is an anomaly. Also found were the teeth of several larger, long-extinct wallabies, but none of the larger living kangaroos, perhaps indicating that there was no grassland nearby.

The Hamilton fauna was dominated by species of the most advanced kangaroo subfamily, the Macropodinae. This is surprising, for in older deposits (from the Miocene period, 5 to 24 million years ago) the more ancient balbarids and bulungamayines predominate. Their absence from Hamilton indicates that by 4.46 million years ago a continent-wide revolution had occurred in Australia's kangaroo fauna, with the macropodines sweeping away most of the competition—even in the ancestral rainforests.

A change in the world's climate may have contributed to the meteoric rise of the macropodines, for around 6 million years ago the Earth entered a dry and hot phase that is marked by the Messinian crisis. This is when the Mediterranean dried out, leaving a massive bed of salt behind, which is still preserved under the sandy bottom of that azure sea. It is also when our first upright ancestors appear in Africa, marking the expansion of drier, more open environments there, and also when many extinctions occur among the mammals of North America.

Changes in Australia's environment at the time of the Messinian crisis may have benefited macropodines in two ways. As the continent dried out, distances between watering points and areas of good feed would have increased, and the hopping mechanism perfected by macropodines would have provided an enormous competitive advantage. And if the drying trend led to the expansion of grasses (as some suspect it did) and other tough-to-digest plants, innovations in the stomach may have helped as well, for by this time the macropodine kangaroos had developed a method of digestion every bit as radical as hopping itself.

Kangaroos, acting alone, are incapable of digesting grass. Instead they harbour bacteria and other organisms in their expansive, sack-like stomachs that chemically attack and break down fibrous plant matter. It is a neat symbiosis, the kangaroo acting as a sort of mobile fermentation vat, while its stomach fauna feed the host with waste products and dead bodies. It provides a striking parallel with the ruminants, whose multi-chambered stomachs are well known. Kangaroos even ruminate (chew the cud) after a sort, though in their case the process is known as 'merycism' and can result in the forceful ejection of a bolus of food from the mouth. While a little more like expectoration, perhaps, it is close enough to dub the kangaroos as a type of marsupial ruminant.

There are, however, significant differences between the digestive systems of cows and kangaroos. The stomach-guests of cows are mostly microscopic—single-celled organisms and bacteria—while those of kangaroos can be grotesquely large. Inside the stomach of a kangaroo that has been feeding on prime quality green pick you will see a mass of wiry worms, as thick as a hairpin and twice as long. Sometimes these creatures are so densely packed that you can't even see the grass the kangaroo has swallowed. Known as strongyles, these worms were long thought to be parasites, and a fine excuse they were too for those agitating against the consumption of kangaroo meat. More detailed studies, however, have revealed that they are the kangaroo's little helpers, ingesting the grass that the kangaroo cannot break down, and feeding the kangaroo with their waste products. Their abundance is thus a sign of a healthy kangaroo.

So are kangaroos best described as herbivores or carnivores? In 1994 I located a tree-kangaroo entirely new to science. Dingiso is the size of a labrador, strikingly marked in black and white (somewhat resembling a panda) and it lives in the alpine scrubs atop the highest mountains in West Papua. When I asked Lani hunters what the creature ate they unhesitatingly said, 'Worms.' I scoffed at this, but upon investigating the stomach of one of the creatures I discovered a knotted mass of more than 250,000 large strongyles, including two species unknown to science, and between which hardly a vestige of the leaves the kangaroo had swallowed could be seen. The Lani were absolutely right, for the kangaroo is host to the worms, which it feeds chewed leaves, while the creature itself feeds upon the worms and their by-products. Thus, in a very real sense, some kangaroos eat worms even though they swallow leaves or grass. It is unusual, incidentally, for tree-kangaroos to have strongyles. This is because they get into the stomach as eggs, from kangaroo faeces carried to the mouth when feeding. Tree-kangaroos do not usually get infected because their faeces drop many metres to the forest floor, breaking the cycle. Dingiso, however, is the only tree-kangaroo to live on the ground, allowing it to benefit from this unusual relationship.

Strongyles are only useful in certain circumstances. If a creature with such stomach-guests feeds on more digestible food the guests will simply beat the host animal to its meal. It is only where they can break down a food that the kangaroo cannot tackle by itself that the relationship benefits the kangaroo. Grass is just such a food, and with the help of their stomach-guests the macropodines have been able to utilise a once unavailable food source. Grass, in effect, became the kangaroo's new frontier.

Was hopping or digestion more important to the success of macropodines? The tree-kangaroos provide an answer. As you might imagine, hopping through the canopy can be dangerous, for one slip means death, and tree-kangaroos can be clumsy creatures. Indeed, so unsuited is hopping to life in the treetops that some species of tree-kangaroo have re-learned to walk when aloft. Despite this, though, the tree-kangaroos have been wildly successful. The ten living species dominate the large herbivore niche in Australasia's rainforests, presumably because their digestive efficiency more than makes up for their clumsiness. In their case the macropodine stomach found a somewhat different use, for their stomach-guests help destroy the toxins that the leaves of rainforest plants are often laced with, as well as helping break down the leaf's cuticle.

It has long been believed that the success of the macropodines was the result of drying around the time of the Messinian crisis. The example of the tree-kangaroos, however, along with the abundance of macropodines in the ancient rainforests of Hamilton, leave me unconvinced. Is it possible, I wonder, that the macropodines arose in rainforests and initially evolved into many types there before spreading into drier environments? As long as the 'ghastly blank' in our fossil record from around 7 to 4 million years ago persists, we are unlikely to be able to answer this fundamental question. We do know, however, that in time the macropodines would reach their finest flowering in the grasslands and woodlands of Australia, and the story of that episode in their history is best read at a fossil locality 1000 kilometres northeast of Hamilton, in the Hunter Valley of New South Wales.

Grass for the Kangaroos

Bow Creek is a small waterway near the town of Merriwa in the Hunter Valley, around 200 kilometres northwest of Sydney. It wends its way through a bucolic landscape of managed pasture and stately river red-gums typical of the western slopes of Australia's Great Dividing Range. A chance discovery by a road worker in the 1960s revealed that the region once had a far richer fauna. He had found the jawbone of an extinct giant rat-kangaroo allied to *Propleopus*, an exciting find which had palaeontologists from the University of Sydney scurrying out to investigate what turned out to be a very rich fossil field.

By the time I arrived on the scene in 1982 the locality had been well worked over and a small collection had been assembled. Reflecting on the fossils and the proximity of the site to Sydney, Michael Archer decided this was a good place to give students a taste of life in the field. The fossils were preserved in an old river 'terrace' about twenty metres above Bow Creek. The terrace was once the creek bed and its winding course can still be traced for kilometres across the countryside.

The bones proved impossible to date, but comparisons with the Hamilton site indicate that the Bow fossils were around 4 million years old. This interested me greatly, for it was a critical period for the evolution of kangaroos.

Road-cuttings can be dangerous places, and the one at Bow Creek was narrow and steep, so with twenty or more enthusiastic students clambering about its sides, the passing of a cattle truck often had me—a tutor responsible for the students—thinking more of liability and death by misadventure than ancient bones. Other difficulties emerged from the Department of Main Roads, which objects to people undermining their road cuts. This, combined with the fact that the owners of a house perched beside the cut had visions of their abode tumbling into the traffic, made our excavations at Bow less extensive than we would have liked.

Despite such impediments, our understanding of the fauna grew each dusk as we gathered around the barbecue at the Merriwa caravan park to spread the day's booty on a picnic table. Then, over a glass or two of the wine the Hunter region was rapidly becoming famous for, we would put our deductive powers to work. What species did this jawbone belong to? Why was this backbone still articulated, while that bone was rounded to a nubbin? The owner of the backbone, we concluded, must have died very close to where its remains were interred, while the nubbin could have come from a creature that lived and died miles upstream. Thus was a picture of Bow Creek as it was 4 million years ago gradually assembled.

Most of the species were entirely new to science, and it became part of my doctoral research to sort out which bones belonged with which set of teeth, and to name the species. Around nine out of every ten bones unearthed at Bow belonged to a member of the kangaroo family, the old river gravels being crammed with the remains of animals that, were they living today, would be called wallabies—a catch-all name for smaller kangaroos.

The fossil 'wallabies' were breathtakingly diverse, the most common belonging to a genus known as *Troposodon*. They are very rare in most fossil deposits and their heyday was around 5 to 3 million years ago. Though five times larger, they are related to the cat-sized banded hare wallaby (*Lagostrophus fasciatus*), which is the last survivor of a diverse and numerous subfamily of kangaroos. Known as the Sthenurinae, it included the troposodons and the short-faced kangaroos, which we will hear more of later.

The banded hare wallaby is a beautiful soft grey, its rump is marked with a series of deep chocolate-and-silver bands. It is now, sadly, a relic of a relic, for not only is it the last of its lineage, but it clings to existence on just two islands in Western Australia's Shark Bay. You can trace its decline on the mainland through specimens lodged in various European museums. The first population to vanish lived in South Australia. Until 2003 it was thought to have become extinct before the arrival of the Europeans, but then I recognised a specimen collected in 1863 in the Gawler area, which had lain unidentified in a glass-fronted cabinet in the Humboldt University museum in Berlin for 140 years. It represented a new subspecies, which my student Kris Helgen and I named.

The South Australian banded hare wallaby was mostly reddish rather than grey, and sported a very handsome and luxuriant black crest on its short tail. We might know more about it had not a second specimen been thrown out in the nineteenth century because it was 'too rotten'. But I suppose that we should be grateful for small blessings, for beside the sole surviving example hung a number of scorched wallaby skins, one of which bore a handwritten label indicating that it had caught fire on 1 May 1945—seven days before war's end—as Russian troops, engaged in hand-to-hand fighting in the museum, tossed a grenade at the retreating Germans. Had that grenade veered a metre to the side we may never have known what South Australia's unique banded hare wallaby looked like.

The western subspecies of the banded hare wallaby fared a little better. Collections in London's Natural History Museum and at Oxford University indicate that it persisted in the forests of the Darling Escarpment inland from Perth until at least 1906, where it may have been protected by a peculiar native plant known as poison peas. *Gastrolobium* is so toxic that even a mouthful can be fatal to a sheep. The active ingredient is fluoroacetate, better known as 1080, and through long exposure to this toxin the marsupials of the southwest have become remarkably immune. By a quirk of its chemical structure, members of the dog family are exquisitely sensitive to fluoroacetate, and the tiniest dose is fatal for a dingo or a fox. So while the poison-pea bush remained dense the little wallaby persisted. Nevertheless, changes in the early twentieth century—to do with land-clearing and fox predation—led to its rapid extinction on the mainland, leaving only a few hundred on Bernier and Dorre islands.

The other 'wallaby' fossils discovered at Bow belonged to creatures that approximated the living swamp wallaby (*Wallabia bicolor*) of the east coast and the agile wallaby (*Notamacropus agilis*) of Australia's north. These animals are browsers (meaning that they eat a variety of foliage, but not much grass) or browser-grazers (which eat more grass, but are not dependent on it). No remnant of the vegetation that fed Bow's kangaroos has survived, but a crocodile tooth preserved nearby suggests that conditions were then warmer than today, perhaps allowing more-tropical plants to thrive.

From the abundant remains of browser-grazers we can surmise that the vegetation of the site was relatively open, perhaps a woodland with some grass and thickets. Two waterworn teeth of a tree-kangaroo testify that in the upper reaches of Bow Creek patches of rainforest persisted, while a few teeth of a euro—the only larger kangaroo found at the site—indicate the presence of rocky slopes and grass. Not a single scrap of the *Propleopus*-like giant rat-kangaroo was ever recovered during our dig. The dozer driver clearly hit the jackpot the day he jumped off his machine to pick up the jaw of this most elusive of fossils.

In effect, Bow Creek marks the modernisation of the kangaroos, for by then almost all of the major genera known today, as well as the immediate ancestors of many of the huge ice-age types, had come into existence. The fossils from Bow Creek demonstrate how the dominant trend in kangaroo evolution has been to colonise ever more open habitats. But as always there is the exception that proves the rule, for one kangaroo has forsaken the plains to return to life in the dense scrub. The creature was known to the Eora people of the Sydney region as *bugaree* (pronounced something like 'buggery'), while the European settlers knew it by the misleading and bland name of swamp wallaby (*Wallabia bicolor*). Although molecular studies reveal that it is closely related to the large red and grey kangaroos and the striped wallabies, its teeth are fundamentally different, indicating a reversion to a life of eating soft herbage. It is a great survivor, for it is found over a wide area of eastern Australia—often far from swamps—wherever vegetation is dense.

The fur of bugaree is a delicious licorice colour, tipped with silver, while its underside is often a deep, rich ochre; yet the beauty of the animal is difficult to appreciate in nature, for it is rarely seen outside dense thickets. But if you are bushwalking in any of Sydney's larger national parks and you go quietly, you might catch a glimpse of one. If it is a warm day the animal might be sunning itself beneath the north face of a sandstone boulder, its eyes half closed in pleasure, its rich, red belly exposed—much in the pose of a sunbather and as relaxed as any Bondi beauty. It will not be easy to spot, for its ochre belly will mirror the iron stains on many rock surfaces. Perhaps a twitch of one of its shortish ears will alert you; or maybe all you will see is the white tip of its long tail disappearing into the bushes as the vigilant bugaree, having seen you approaching long before you saw it, flees for safety.

Bugaree has a paltry number of chromosomes: eleven in the male and ten in the female—the smallest number of any marsupial. It nevertheless uses its genetic inheritance to great effect, for it is able to eat toxic plants—even introduced ones such as the hemlock that saw off Socrates—with no evident ill-effect. But it is the tweak that it has given to the kangaroo method of reproduction that is its most astonishing achievement.

Unlike other kangaroos bugaree's females do not have to await the birth of their young in order to mate and conceive again. Instead she mates up to eight days before the birth, thus gaining an extra week to grow her embryo before it is thrust out into the world. No other marsupial can do this, for none, except some tree-kangaroos, can maintain a pregnancy for more than the length of their oestrous cycle. Just how the animal manages this in terms of hormones, immune responses and such like, is not known. But bugaree must have many secrets that we have not even begun to guess at, for it is the last of a once numerous tribe of kangaroos to survive around Australia's eastcoast cities. If you live in Melbourne, Sydney or Brisbane, it's a fair bet that bugaree is closer than you think. In Sydney it still appears as close to the city centre as the Manly reservoir, just twelve kilometres from the Opera House, while around Melbourne the Dandenongs and Cranbourne areas remain favoured haunts. In his memoirs written around 1883, Sydneysider Obed West recalled that they were 'very numerous' in the vicinity of what is now the Australian Museum on College Street, and 'many scores of them have I shot' in a wetland just south of the museum site. Once upon a time potoroos (another Eora word), bettongs, rufous rat-kangaroos, rock-wallabies, red-necked wallabies and eastern grey kangaroos all thrived around Sydney. And there is no reason why these creatures should not flourish once more, for Sydney is blessed with parks and reserves, and kangaroos are adaptable creatures. All they require is a little assistance from the city's humans.

Every year, usually around June, the first winter winds blow in from the west over Sydney, chilling the estuaries and wetting the vegetation. It's a hard time for bugaree, and it is then that you might notice something furry floating in the waters of Botany Bay or the Hawkesbury estuary, carried by the tide towards the open sea. If such a sight passes you by as you sit fishing on a jetty or pontoon, you might wish to spend a moment of two examining these last mortal remains of bugaree. You will find that her teeth are worn to the gums and her claws blunted by years of clambering over rough sandstone boulders. She is old—very old—and has not fed well for months now, for her swollen joints make travel painful and her worn teeth cannot chew the plants that give her sustenance. But still she breeds, the long nipple in her pouch indicating that right up until the last she suckled a joey. She is a survivor—perhaps she too has 'the will to fail'—and she commands our salute as she goes to her final rest.

16

Not Formed for Such Work

On Sunday 12 October 1872, the explorer William Hann was in the rain-forest wilds of northeast Queensland and, judging from his diary entries, the humidity and rugged topography were leading to a shortness of temper. The last straw, it seems, was a story told by his Aboriginal guides of a kangaroo that lived in trees. This 'fable' was, to Hann's chagrin, taken seriously by the expedition's doctor, Thomas Tate, who claimed to have located a 'nearly complete skeleton', with which he wanted to burden the expedition horses.

At this imposition Hann's temper broke. He 'demurred at my taking such rubbish' the miffed Tate noted; while Hann wrote in his diary, 'To entertain the idea that any kangaroo known to us, or approaching its formation, could climb a tree, would be ridiculous; the animal was not formed for such work.'

But kangaroos in the treetops there are. Eleven years after Hann tossed out Dr Tate's skeleton, a Norwegian naturalist by the name of Carl Lumholtz collected, in those same scrubs (as rainforests were then called), a complete specimen that set the scientific world aflame. The extraordinary discovery was named in his honour and has been known as Lumholtz's tree-kangaroo (*Dendrolagus lumholtzi*) ever since. When, a few years later, the Reverend Charles De Vis of the Queensland Museum obtained a specimen of a second, larger species from further north, the scientific world was at first suspicious—surely this was an aberrant Lumholtz's tree-kangaroo. Bennett's tree-kangaroo (*Dendrolagus bennettianus*) was, however, a valid species, meaning that Australia's tropics were home to not one but two species of kangaroos living in the treetops.

Astonishment at the discovery was largely limited to the English-speaking world, for ever since Dutch scientists from the Netherlands East Indies Natural History Commission had found tree-kangaroos in New Guinea in 1828, Europe had known of such creatures. But to the Anglophone nations (and especially the colonial Australians) Lumholtz's find was novel enough to precipitate a fascination that has persisted ever since. In truth the centre of tree-kangaroo diversity is New Guinea, where eight species reside. The two Australian species are rather primitive and certainly not as accomplished at climbing as some New Guinean types. But like all tree-kangaroos they are (except for a few populations of Lumholtz's tree-kangaroo) reclusive and hard to study.

My sole opportunity to observe Australian tree-kangaroos in the wild came courtesy of Roger Martin of Monash University. In the early 1990s he was undertaking a long-term study of Bennett's tree-kangaroo, then one of the most elusive and poorly understood of all kangaroo species. Roger had camped at a place called Shiptons Flat, inland from Cooktown, and invited me to join him. He had provided me with instructions scrawled on a scrap of paper indicating how to find his camp, and so I drove out of Cooktown, past the Lion's Den Hotel (where Roger occasionally consoled himself), and up a bush track, where the vegetation began to thicken. I was nearing the X-marking-the-spot on the grubby paper when, passing through some thick bush, I came across a large truck parked in the middle of the track. There was no easy way around this obstruction, so I stopped to examine the situation. It was a near-new cattle truck which had suffered considerable damage, and had what looked like a huge pool of oil under it. I bent down and dipped a finger into the sticky substance. It was a pool of blood, at least three metres long and a metre wide.

The hair started to rise on the back of my neck as I wondered what atrocity had been committed on this loneliest of roads. No quick escape was possible, for the nearest help was at Roger's camp, and to reach it I needed to drive around the truck. I quickly marked a path through the dense bush and back onto the track. Over the next rise Roger's camp-fire and chair came into view, though, as I drove on I could see the camp had been utterly trashed. Roger's equipment had been scattered, the tarpaulin he slept under flung into the bushes. Fearing that the murderers at the truck—whoever they were—had done away with my colleague and mate, I was desperately searching for a deceased Roger when I heard a voice behind me.

'What the fuck's going on here?' It was Roger, radio-tracking antenna in hand, puffing up the hill after a morning spent in pursuit of his tree-kangaroos.

'You tell me!' I blurted. 'God, I'm bloody glad to see you.'

We drove to the homestead of the Roberts family, pioneers of the Cooktown district, to find out what was going on, and I told Roger about the sinister truck.

The Roberts' house was a fantastic piece of antique Queenslandiana, its water tank crafted out of the trunk of a colossal tree, and before we reached the front gate we were greeted by Charlie Roberts. The original Roberts had come from England's west country three generations earlier, but so profound was the family's isolation that they still spoke with a marked accent. Added to Charlie's peculiarity of repeating things, it made for interesting conversation.

'Well, those fellas from Wudjal are setting up a butchery—needed some beasts they did,' he began. 'Yes, needed some beasts. Got a new truck with a government grant, a brand-new truck with a government grant it was. Loaded 'em up just here above your camp, Roger. Loaded 'em up, and put their best rodeo rider in the driver's seat, too. I said to him, "If you pump the brakes like that, they're likely to give way sudden-like." Likely to give way, I said, and just then she started to creep forward. Had her in neutral he did, and she started to creep. You know that bend just above your camp, Roger? Well, that fella hung on to there. Best rodeo rider they had too, he was, but he jumped out as she veered into the turn.

'The truck hit that big blue-gum, Roger, that big blue-gum near your camp. Sheared the engine mounts they did and flattened your camp. They tried to drive on but she was so crook they only got to the bottom of the hill and those steers had broken legs and cuts all over, so they butchered 'em on the spot. Left the truck right there, too, Roger. Reckon I might try to buy it off 'em. Might buy it off 'em and fix it up, I might.'

Reflecting perhaps on the tyre mark over his swag, Roger's singular reply was, 'Good thing I got an early start this morning, Charlie.'

The local Aboriginal community at Wudjal Wudjal south of Cooktown includes descendants of the people who gave the word *kangaroo* to the English language. Old Harry Shipton, then in his eighties, was the elder of the clan. He had hunted tree-kangaroos with dogs at Shiptons Flat right into the 1960s, but of late he had found more congenial occupations, for many a young woman had travelled north to spend time in the hippy communes dotted around the region, and oftentimes they sought out Harry to learn a little of his 'blackfella knowledge'. Old Harry was only too happy to share, taking them to remote beaches and forest camps where, according to local gossip, a rather full and intimate knowledge was imparted.

Roger, however, had found it difficult to get much information about tree-kangaroos—or anything at all—out of the old man. Perhaps he lacked the right currency. Indeed Roger's best insight came from a Hungarian emigre who seemed to have a can of XXXX welded to his hand, and was often found sitting out the front of the Lion's Den. One afternoon, after months without seeing a single tree-kangaroo, he asked Roger what he was doing in the district. When Roger explained his interest in tree-kangaroos, the man said, 'Plenty out the back,' jerking his thumb towards the woodland and thin forest corridor that clung to the Annan River just below the hotel. Roger thought finding tree-kangaroos there about as likely as finding fairies at the bottom of his garden, so neglected to investigate. Years later, however, he learned that tree-kangaroos abounded along the Annan. Had he followed the sage advice he could have done his work in half the time from the comfort of the pub itself!

Bennett's tree-kangaroo has rust-red shoulders, a black belly and a long, bicoloured and tufted tail, like a lion's. Charlie told us that when his dad had assisted an expedition from the American Museum of Natural History around 1948, the creature was scarcely seen in the area. Apart from a couple of nineteenth-century accounts and a few museum specimens it was almost unknown, the only recent sightings being from the highest peaks in the region such as the evocatively named Mount Misery. Things remained that way until the 1970s when, Charlie said, about a decade after the last of the old Aboriginal hunters stopped coming to the area to hunt, the population began to grow.

In the 1960s Charlie had seen one hunter with a catch of seven tree-roos, and believed that hunting of the slow-reproducing creatures had kept numbers low. They had survived on the mountain-tops, he thought, because the peaks were story places—haunted regions where the Wudjal people feared to go. Lately, however, they were being seen in entirely new places. One had even jumped through the Roberts' kitchen window, lured by ripening bananas and frightening the bejesus out of the whole family.

After putting the camp in order we set off to track the tree-kangaroos that Roger had collared with a transmitter. Roger led the way through the thick scrub, the strength of the 'pings' from his radio receiver revealing the proximity and direction of the kangaroos. As we walked Roger explained how in order to fit the transmitter he had had to tranquillise the animal and then catch it in a net. Occasionally they would be disturbed and jump, thereafter Charlie would give chase and, astonishingly, he managed to run a couple of them down.

Eventually we found ourselves standing beneath a forest giant festooned with vines and epiphytic ferns, the signal from Roger's receiver indicating that high above lurked a mother tree-kangaroo and her half-grown young. We could see nothing of them, but after half an hour of craning our necks Roger caught sight of two long tufted tails. They looked like the broken ends of vines, and to this day they are all I have seen of this extraordinarily beautiful kangaroo.

Roger suggested that on my way back south I should visit a man who had looked after an orphaned joey Bennett's tree-roo. Rob Whiston was a wild Irish transplant who had created a paradise by the Bloomfield River. Native fish by the hundred came to the creek below the house each day for Rob to feed them, and the verandah was ornamented with a roost of tiny bats. As we toured his estate he spoke of his orphaned tree-kangaroo, a little female that, when annoyed, would take a hop forward with her razor-sharp claws raised high above her head. 'Ready to strike,' he explained, 'but always silently, and with the face as devoid of expression as only tree-kangaroos and the English are capable.'

Rob had planted a hectare of durian trees up the back and was hoping that their fruit would one day provide a living for him and his wife. I mentioned to Rob that in Indonesia orang-utans love durian fruit and strip the trees quickly. Without batting an eye the Irishman said, 'Oh, you needn't worry about them. We'll spray for them, to be sure.' To this day I'm not sure whether he was joking or not.

As I drove off I considered in some detail the similarities between tree-kangaroos and the great apes. One New Guinean species, the Doria's tree-kangaroo (*Dendrolagus dorianus*) is rather gorilla-like in both colour and behaviour, for it eats leaves and lives in male-dominated groups, the alpha males of which are muscular and forbidding creatures. Like silverback gorillas, male Doria's tree-roos use brute force to keep other males from their harem, which saves their sperm from having to compete to fertilise the egg. As a consequence of this, they have small testicles for their body-size. Male chimpanzees, on the other hand, are not as physically forbidding, and because several males co-exist in the one troop they compete sexually as well as physically. As a result they have huge testicles which produce enough sperm to swamp the competition.

Strangely enough, there is a tree-kangaroo that provides a parallel with the chimps, and perhaps even us. Goodfellow's tree-kangaroo

(*Dendrolagus goodfellowi*) is a reddish-coloured fruit-eater and although it weighs only seven kilograms its testicles are as large as a human's. There are even indications that it may form pair bonds, as on occasion do chimps and, most famously, ourselves. In many ways tree-kangaroos are the greatest untapped vein in kangaroo research for, like the first kangaroo, they have crossed into a new frontier and therein have diversified spectacularly.

After nearly twenty years of field work, during which I had the privilege of naming four types of tree-kangaroo, I felt that I had learned most of what I ever would about these fascinating creatures and their evolution. But science has a way of surprising you, and I could not have been more amazed when one afternoon in August 2003 I entered the Western Australian Museum in Perth to look at fossils from caves on the Nullarbor Plain and found, among the clay-smeared bones, the skeletons of two tree-kangaroos. Nullarbor means 'no tree' in Latin and the name is remarkably apposite. So what were tree-kangaroos doing on the no-tree plain? Preliminary dating suggests that they are perhaps a million years old, and in that bygone era maybe there were trees on the Nullarbor. The skeletons are not as adapted to an arboreal life as the living species, yet their tree-climbing proclivities, as revealed by their hind-limbs, are unmistakable. Their skulls, however, surprised me, for they bore similarities both to tree-kangaroos and the pint-sized and strictly terrestrial quokka of Rottnest Island. There is enough of a similarity about those Nullarbor skulls to have me wondering if the quokka is a tree-kangaroo that, during the course of its evolution, has come to ground. Studies of the quokka's DNA, which are yet to be done conclusively, could prove whether such notions are fanciful or not.

1 7

Land of Giants

In 1873 the famous nineteenth-century anatomist Sir Richard Owen, then director of the British Museum of Natural History in London, described a massive jawbone that had been unearthed beside the Tambo River in eastern Victoria. Although the creature would have been the size of an ox, Owen was convinced it had once belonged to a kangaroo and had accordingly named it *Palorchestes azeal*, meaning 'ancient leaper'.

Few dared query such an authority and around 1958 Harold Fletcher, the curator of fossils at the Australian Museum in Sydney, decided to give the public some idea of what *Palorchestes* looked like. Taking a grey kangaroo for his model, he had a commanding reconstruction made. Towering more than three metres, the monstrous plaster figure was for a short time a great drawcard and object of awe. Yet it was fated to come to a bad end, for a few months after it went on display, Jack Woods, director of the Queensland Museum, took a closer look at the jaw of *Palorchestes* and discovered that it lacked the distinctive hole through which the chewing muscles pass in all kangaroos.

Palorchestes was in fact a distant cousin of the wombat, a revelation that so filled the museum staff with embarrassment that they took to their creation with a sledgehammer. Moral: don't believe everything you see in a museum—or on TV, for that matter. Yet giant kangaroos did once bound over Australia's inland plains, and though none were as large as the Australian Museum's fantastic plaster model, some were far more unusual.

Australia's ice-age kangaroos have fascinated me ever since that first opportunity in 1974 to clean and study their skeletons as a museum volunteer. In 1978 a special chance to study them arose when Tom Rich found himself with so many volunteers at Hamilton that there was insufficient room for all to work. To ease the congestion, Tom gave permission for two to accompany me to a fossil deposit I had located near Minhamite, eighty kilometres east of Hamilton. There, fossilised bones are preserved in a deposit of thick black clay on the bank of a small creek. Although it was 'only' tens of thousands rather than millions of years old, I was excited by the deposits because they contained the remains of many gigantic kangaroos.

To sit in the mud on a frigid summer day such as Victoria's Western District can offer, slicing through stiff clay with a trowel and encountering massive mahogany-coloured bones, became the most exhilarating experience of my life—something I could only be driven from by horizontal sleet, thence to huddle in the back of a panel van to warm myself as the slurry swirled all about. To flick a triangle of clay from such a bone, revealing its shape and knowing that you are the first living being in 50,000 years to see that sight, was as energising as sex; for then you could visualise the part of the skeleton that the bone represented, and fantasise about the kind of marsupial from which it came.

Sometimes, intimate insights into the life and death of an animal would become evident: the slice-mark made by the tooth of a marsupial lion (imagine seeing the groove where that great slavering premolar found its mark), or an old fracture that had healed—each one as informative as the scar on the body of a lover. But what held me most in thrall was trying to imagine those animals as they really were. A huge grey kangaroo, almost twice the weight and a third taller than any living today had stood on this spot. Did it endure the sleet as I did? Did its nostrils flare at the scent of its mate; and how did it breathe its last, right here, in the black clay under my feet?

I was holding its ankle-bone in my hand 50,000 years after its burial, a still-living thing on its land, yet separated from its life by such a gulf of extinction and change as might separate me from my unimaginably distant descendants, if humanity and my genes survive so long. The gulf of time will consume you if you linger over it too long—it will break down your morality and your essence, so that you, like the extinct kangaroo, will only fuck and eat and sleep, until you too join the black mud.

And what of the extinct giant wallaby whose empty eye-socket looked out at me as I excavated its skull a short time later? What land did it behold, and what were its comforts? We need not only more scientists but poets as well—a Ted Hughes of palaeontology—who can imagine those past lives and guide us through the labyrinth of time to show us how things were in the distant past.

The story of the Australian ice age—the most glorious age of kangaroos—was first glimpsed in caves located a few kilometres east of Wellington, New South Wales, where the western foothills of the Great Dividing Range give way to featureless plains. As with Hamilton and Bow Creek, the Wellington Caves lie in a region of undulating hills and red-gum-dotted grasslands. Their secret was first penetrated around 1829 when a settler named George Rankin lowered himself into a shaft on a long rope that he had attached to a protrusion on the cave wall. When it unexpectedly snapped off he discovered it to be the limestone-encrusted leg-bone of a huge, extinct bird. It was the first significant ice-age fossil ever found in Australia, and it led to explorations that revealed the caves to be chock-a-block with the bones of extinct creatures. So abundant are these fossils that they were once mined for their phosphate content, a process that saw millions crushed to dust and strewn on the fields as fertiliser. For all the destruction this entailed, a huge number of fossils made their way into museum collections.

A large collection is held in the Australian Museum in Sydney, and one day in 1981 I found myself examining this treasure-trove with an American scientist. Fred Szalay, of the City University of New York, has a stellar reputation in the field of mammalian evolution, and as a young student I was honoured, even a little intimidated, to be in his company.

Fred and I had both come to the conclusion that the evolution of various animal groups could best be approached via the study of their feet. This may sound unorthodox, but for creatures with such a strange method of locomotion as the kangaroos it felt natural to focus on feet for enlightenment. Fred had spent decades studying the feet of everything from opossums to pangolins. He taught me that a single foot-bone is, in many cases, sufficient to identify a species—it's astonishing how distinctive the foot-bones of otherwise similar-looking creatures can be. On that day we sorted the ankle-bones into piles—grey kangaroo, giant wallaby, short-faced kangaroo—when Fred unexpectedly stopped, a peculiar ankle-bone in his hand.

It was, he said, unmistakably that of a large tree-dwelling kangaroo. Did I know of such a creature? A glance was enough to confirm his diagnosis, and to see that this was no normal tree-kangaroo, for it was twice the size of the largest known species. It is sometimes said that an extinct animal can be reconstructed from a single bone, but in truth, often only broad outlines of the species' ecology can be established. In the case of the owner of the ankle-bone, various facets revealed that the foot could be twisted inwards so that the soles of the feet could oppose each other. Such are the stresses on the foot of terrestrial kangaroos that twisting would lead to a crippling injury, but in tree-kangaroos it is essential, for without the ability to grasp a tree-trunk between its feet it could not climb.

I can only assume that *Bohra*, as we called the creature, lived in relict rainforest thickets that grew around Wellington Caves a million or more years ago, though nothing else suggestive of a rainforest surfaced among the thousands of bones we examined. *Bohra* was a rare creature, for nothing more than a few leg- and foot-bones—all from the Wellington Caves—have ever been recovered. Judging from them, it must have weighed thirty to forty kilograms—as much as a female orang-utan—and was doubtless quite a sight in the treetops.

Bohra is only one rare species among dozens of kangaroos that became extinct in Australia towards the end of the ice age. Had you visited the continent then you might not have noticed the red and grey kangaroos that are so eye-catching today, for larger and more striking species then existed, among them a multitude of short-faced kinds. These creatures had short and thick tails, pot-bellies, short necks, a single toe on each hindfoot (as opposed to the four in most living kangaroos), and blunt, deep muzzles. Their teeth indicate that most were leaf-eaters, and their arms were specialised into great hooks, with two very long digits tapering to long, straight claws which were probably used as rakes to draw foliage to the mouth.

No fewer than thirty species of these kangaroos have been recorded, and they form the dominant fossils in many ice-age deposits. Remarkably, the lineage seems to have come out of nowhere, for only a few fragments of teeth have been found in Hamilton-age deposits; yet by a million years ago they had become *the* dominant type of kangaroo over most of southern Australia. Their flowering, however prolific, was brief, for they vanished even more rapidly than they appeared.

Short-faced kangaroos were found in almost all habitats, from desert to tall forest, and up to half-a-dozen species could co-exist. The smallest were the size of a female grey kangaroo, while the largest were the largest kangaroos that ever lived. When I look at caves filled with their fossilised bones, and at their leaf-slicing teeth, I cannot help but wonder at the land that fed them. Today, over much of the country where they were once found, grass predominates, a food unsuitable for them. True, there are huge areas of saltbush and bluebush in southern Australia that could have provided suitable fodder, but the short-faced kangaroos were far more widespread than the saltbush plains. Southern Australia must have changed in ways that we are yet to fully fathom since their time. It is as if an entire biome has vanished.

The true giants of the short-faced kangaroo tribe were not, according to their teeth, leaf-eaters. Known as *Procoptodon*, these short-faced monsters are a specialty of, as well as a convenient marker for, the Pleistocene period some 2.4 million to 10,000 years ago. While completing this book, Kris Helgen and I devised a simple and elegant method to estimate the weight of these vanished giants. Our preliminary analysis suggests that kangaroos are so proportioned that their weight can be gauged from the dimensions of their femur—the bone of the upper leg. If you measure its circumference in millimetres, then subtract 30, you now have an approximation of the long-vanished creature's weight in

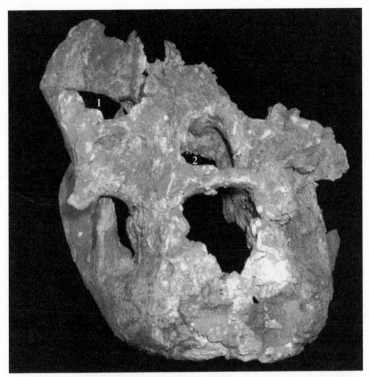

An ice-age mystery—the skull of *Procoptodon goliah*. Its similarity to a human skull is unmistakable. (1 eye socket; 2 nasal cavity.)

kilograms. From this method, we infer that the largest kangaroo that ever lived—a short-faced species known as *Procoptodon goliah*—weighed around 130 kilograms. This is a little less than previous 'guesstimates', and I suspect that as our study unfolds we will find that the weight of many extinct kangaroo species has been overestimated—an important point to keep in mind when we try to imagine the fauna and environment of Australia's ice age.

Judging from the bones that support it, *Procoptodon*'s belly was of Friar Tuck-like proportions. Its fore-limbs were exceptionally long and powerful, with rake-like claws on two of its digits that were longer than human fingers. Its neck was short and its head, with its ridiculously

The largest kangaroo, *Procoptodon goliah*, as it may have appeared in life. Its short face, elongated claws on the hands and hoof-like claw are unlike anything seen in living kangaroos. It became extinct around 46,000 years ago, about the time humans arrived in Australia.

shortened muzzle and heavy jaws, almost box-like. Standing well over two metres tall, *Procoptodon* would have been an impressive sight. But what would you have seen had you looked into that strange face?

From what we can reconstruct, you might have discerned something similar to a gorilla, a lemur, or even yourself—a resemblance wrought by binocular eyes, a distinct chin, and short nose. Overall, the bones are striking in their hominid resonance. The teeth also have an eerie human quality; the molars are heavy and wrinkled, like those of the extinct African man-ape *Australopithecus*, and the enamel is extremely hard. Perhaps the diet of *Procoptodon* was similar to that of the robust australopithecines, consisting of seeds, roots and a little grass. But mystery still surrounds these beasts, for there are so few palaeontologists in Australia that detailed studies of their teeth and the chemistry of their bones (to reveal their diet) are yet to be undertaken.

A second ubiquitous but now vanished ice-age group were the giant wallabies. Belonging to the genus *Protemnodon* they were more widespread than the short-faced kangaroos but less flamboyant in their evolutionary branchings. By a million years ago their heyday had passed, and just three species survived in Australia. The wetter regions were home to a graceful giant known as *Protemnodon anak*. Named for a giant of the Bible, it was around twice the bulk of a living grey kangaroo. It resembled an overgrown swamp wallaby, though its arms were large and heavy, and its neck, at over twice the length of any living kangaroo, made it the nearest approximation to a marsupial giraffe that ever existed. Few details of its ecology are available, which is a shame because exceptionally well-preserved remains including stomach contents and skin impressions have been available for over twenty years, yet remain unstudied.

The inland and parts of the coast were home to two small-bodied but very large-headed *Protemnodon* species. Little is known of them, but it is possible that one had a specialised nose, for its nasal bones are pulled backward in a manner similar to that of a tapir, though not so extreme. The most extraordinary *Protemnodon* of all, however, inhabited New Guinea. The ice-age megafauna of that rainforest-covered island included three species of *Protemnodon* (the only ice-age megafaunal kangaroos known from the island) that were distributed from lowlands to alpine mountain summits.

In 1983 I was fortunate enough to name the first of these to be discovered, calling it *Protemnodon tumbuna* in honour of the first New Guineans (*tumbuna* meaning 'ancestors' in Melanesian pidgin). Then it was known from a single jawbone and a few loose teeth excavated from a cave in Chimbu Province, but around 1990 a near-complete skeleton of this grey-kangaroo-sized creature turned up in a swamp near Tari in the Southern Highlands.

What it revealed transformed scientific opinion, for its head was large in relation to its body, its hind legs absurdly short and stumpy for a kangaroo, and its fore-limbs very powerful. Details of its shoulder and lower arm revealed that its fore-limbs functioned differently from other kangaroos, and had been modified to assist in bandicoot-like bounding. *Protemnodon tumbuna,* it seems, was the only kangaroo ever to relinquish that invaluable gift of its ancestors—the ability to hop. Instead it made its way through New Guinea's dense rainforests on all fours, rather like a bandicoot or rabbit.

Although the protemnodons are extinct, the rainforests of New Guinea are still home to a generalised, dwarfed version of these creatures. Known as forest wallabies, *Dorcopsis* (meaning 'gazelle-faced') still abound in lowland rainforests throughout the island. There are three species, which replace each other in a ring pattern around the

Protemnodon anak (upright), a giant wallaby that until around 46,000 years ago inhabited eastern Australia. It was the closest thing the kangaroo family ever produced to a giraffe. *Protemnodon tumbuna* inhabited New Guinea's mountains and was the only kangaroo ever to relinquish hopping.

lowlands: one with a bold white stripe down its back in the north, a yellowish and ash-coloured one in the southeast and a chocolate-coloured one in the southwest. Around the size of a swamp wallaby, they all have fine, silken fur and long faces. The powerful arms of the males (three times as big as the females) may be used to 'wrestle' for mating opportunities. But it is their teeth that are most unmistakably *Protemnodon*-like, for their premolars are two-centimetre-long blades, while the molars are low-crowned and entirely unsuited to eating grass. Scientists suspect that *Dorcopsis* wallabies use their premolars like scissors to cut up leaves and fruit, but some New Guineans think otherwise. In 1984 I asked a Mianmin hunter from the Upper Sepik region what the forest wallabies ate. He insisted that they consumed insects, found by turning over stones on riverbanks.

The behaviour of these wallabies in the wild is almost entirely unstudied, which is a great pity, for they are very odd creatures. Their tail is never laid flat on the ground, but carried in an arch, with just its tip making contact with the earth. It looks like a cumbersome way of getting about, and no one knows quite why they do it, but it has been suggested that leeches are so plentiful in the rainforest that they could suck the blood from a wallaby in no time if it were to lay its tail on the ground. The tail has a cornified tip which may be very useful in pushing the animal along when held in an arch. Interestingly, some *Protemnodons* appear to have lacked this specialisation.

Only one more extinct genus remains in this catalogue of giant ice-age kangaroos—*Propleopus*—and here we have no living species to guide us in our thinking. This is an extremely ancient lineage—I would argue the most ancient of all, and was probably already separate by Lake Pinpa times, 20 to 40 million years ago. For creatures with such a long tenure on our planet they have left us far too few remains—a tooth here and a jaw fragment there—with no site, no matter how rich in fossils, yielding more than a handful of pieces. In 1981 as part of my doctoral studies, I gathered together most of the known fragments, which fitted into a shoebox packaging and all, to see what I could learn about them.

There were three ice-age species which although similar in size (around forty kilograms), had rather different teeth. The most widespread species, *Propleopus oscillans*, once occupied much of southeastern Australia, from Queensland's Darling Downs to the southeast of South Australia. It had a very stout and sharp, almost dagger-like, forward-pointing lower incisor, four crushing molars, and a modest-sized buzz-saw-shaped premolar. A second species, *Propleopus wellingtonensis*, which is known only from a single jaw found in the Wellington Caves, had somewhat larger premolars; while the third, *Propleopus chillagoensis*, whose remains were found in the Chillagoe Caves of northeastern Queensland, had massive, serrated premolars and rear molars reduced to rounded bumps.

I wanted to know how these creatures lived. With only teeth and jaws as testimony to their existence, I thought that I should start with diet. After all, you are what you eat. The answer at first seemed straightforward. The musky rat-kangaroo of Queensland's tropical rainforests has similar teeth and eats seeds, fruit and insects. Perhaps, I reasoned, this is what *Propleopus* did.

But the bones of *Propleopus* were often associated with creatures that inhabited woodlands and plains. In such places fruit and nuts are uncommon, certainly not existing in sufficient quantity to sustain a forty-kilogram kangaroo. And the idea that any creature of that size could find enough tough-skinned insects (such as beetles) to satisfy it is ludicrous. As I pondered these conundrums I wondered why the remains of *Propleopus* were so widely distributed, yet so rare that an avid fossil hunter like myself should consider it 'probably hopeless' that I'd ever unearth one.

I decided to go straight to the teeth themselves to determine if the food they had once bitten had left any traces. Almost as soon as I loaded the first *Propleopus* premolar onto the platform of a scanning electron microscope the answer leaped out. The worn surface of the blade was gouged by deep, parallel grooves. The teeth of just two other marsupial species—the thylacine and the marsupial lion—show similar marks. Tooth enamel is tough stuff, and the exteriors of neither nuts nor insects could do such damage. You need a mineral to gouge such grooves in tooth enamel, and in this case the mineral was probably quartz, in the form of sand grains that had been caught in the coats of the victims of flesh-eaters.

Suddenly the mystery of the *Propleopus* made sense. Those stout incisors had been used to stab flesh and the buzz-saw-shaped premolars to slice it. And of course predators are always fewer in number than their prey, explaining the rarity of the creatures in the fossil record. But was it really possible that carnivorous kangaroos once roamed Australia?

Although the evidence fitted the theory splendidly, at first I resisted the idea of a carnivorous kangaroo. On reflection, however, I realised that herbivorous lineages have on occasion given rise to carnivores; the marsupial lion is a carnivorous wombat-relative, while us humans are carnivorous descendants of herbivorous apes.

Then there were a delicious couple of days when, as I worked on my theory without telling anyone else, I was the only person on Earth who knew that great, carnivorous kangaroos once stalked Australia. The hypertrophied premolars and reduced molars of *Propleopus chillagoensis* indicate that it was the most devoted flesh-eater of the trio, while its more widespread cousin *Propleopus oscillans* was, judging from its large

The killer kangaroo, *Propleopus oscillans*, was a meat-eater that once roamed much of eastern Australia.

molars and smaller premolars, an omnivore. Long after I wrote up these findings with my doctoral supervisor, Michael Archer, even more intriguing pieces of these mysterious kangaroos were discovered, including a near-complete skull of a smaller, related Riversleigh species which had a high sagittal crest atop the skull that anchored powerful biting muscles; and a partial skull of *Propleopus oscillans* from a cave in South Australia, with a fine, dagger-like upper canine—all the better to bite you with, we might say of this wolf in kangaroo's clothing.

Despite these breakthroughs, the likelihood of a more complete understanding of *Propleopus* seems as distant as ever. The only bone from behind the skull that has ever been found is a humerus, or upper arm bone of *Propleopus oscillans*, and it is only attributed to the species because, in its resemblance to a human arm bone, it is too weird to belong to anything else. Beyond that, all we have is mystery.

So what can I tell you of *Propleopus* by way of summary? It was around the size of a female grey kangaroo and I doubt that it hopped, for the lineage diverged too early from the kangaroo mainstream to benefit from that innovation. Instead, it probably bunny-hopped like a bandicoot or rabbit. Judging from the large attachment points on the skull, the head was muscular—somewhat in the style of a Tasmanian devil. Its body could have been lithe—designed to run down the likes of rat-kangaroos, birds and bandicoots—or may have been stout, muscle-bound and tail-less, a creature given to tearing into the bodies of dead and dying ice-age giants. Until we find a skeleton, we cannot know more.

1 8

Is the Answer 46?

Ever since George Rankin's downfall led to the discovery of Australia's megafauna, scientists have speculated on what made these animals extinct. Once there were giants. Now there are none. When I wrote an ecological history of Australasia (*The Future Eaters*) in 1994, I tried to determine what might have happened to the giant kangaroos and other megafauna, and to imagine the consequences of their extinction on the environment as a whole. Then, the data needed to solve the mystery was limited, and I had to work from first principles rather than the fossil record. My study pointed towards the megafauna having been hunted quickly to extinction by humans, and that the extinction event led to a long series of dramatic consequences for the entire Australian environment.

The first of those consequences, I hypothesised, was that as most large herbivores became extinct, and the survivors were reduced to low numbers, much uneaten plant-matter accumulated, providing fuel for more frequent and intense fires. This changed fire regime in turn dramatically altered Australia's vegetation, allowing the eucalypts and other fire-promoting species to spread from their original habitat in the regions of poorer soil until they occupied most of the continent. Their expansion came at the expense of 'dry' rainforest and fire-sensitive scrub, which had supported a considerable element of the megafauna, but which after their demise declined to near extinction.

These changes in vegetation led to a dramatic shift in Australia's climate that, American climatologist Gifford Miller pointed out, occurred because less moisture is transpired into the atmosphere from eucalypts than from rainforest trees. Before the extinctions, this moisture was carried inland on prevailing winds, enhancing rainfall by up to 60 per cent.

My hypothesis caused outrage in some sectors of academe, with a fierce reaction coming from those who believed that indigenous people always lived in harmony with nature. The hypothesis was then eminently debatable, for in 1994 we had no reliable dates on when Australia's megafauna became extinct, and only a slim grasp of when people had arrived in Australia. Furthermore, although palaeobotanists had established that a great change had indeed occurred in Australia's vegetation, we had no idea when this momentous shift had taken place.

The controversy was made worse by a series of misleading dates (now mostly discredited) which suggested that Australia's megafauna survived until 6000 years ago, that the first Aborigines arrived 116,000 years ago, and that the vegetation change had occurred as recently as 38,000 years ago—or perhaps as long as 120,000 years ago. This absence of firm dates left me determined to pursue and test my hypothesis by dating the megafaunal extinction event. Fortunately, new techniques were becoming available that would allow us to reach back in time as never before.

In pursuit of this goal I found myself loitering, one chilly morning in July 1997, beside a small crevice located deep in the Margaret River district in Australia's southwestern corner. Beside me stood three other scientists: Gavin Prideaux, then a doctoral student but now the world authority on the short-faced kangaroos; Melbourne University's Dr Bert Roberts, an expert practitioner of a new dating technique known as optically stimulated luminescence (OSL); and the unforgettable Professor Rhys Jones of the Australian National University, a leader in the field of archaeology and one of the most innovative thinkers Australia has ever produced. Gavin explained that the crack before us was the opening to Tight Entrance Cave, a subterranean labyrinth that contained a chamber filled with ancient bones. He had excavated thousands of samples from the cavern, including the jaw of an ancient rat-kangaroo related to the bettongs but twice the size of any living species. This new discovery, he announced with a slight smile, would be given the generic name of *Virginia*, but eventually he opted for *Borungaboodie*, meaning 'very large ground-rat' in an Aboriginal language.

I was keen to examine this fossil deposit first-hand, but had recently been trapped in a cave in West Papua, and tight squeezes underground filled me with horror. I tried to lower myself in, but choked as I felt the limestone walls grip me. I couldn't do it, and backed out. Rhys Jones, who was as round as a butterball, grasped the rope and forced himself into the gap. What none of us knew then was that Rhys's shape was partly the result of a liver swollen with chronic myeloid leukaemia, a disease that a few years later would take him from us. He had been feeling unwell for months, yet on that morning the Welshman displayed such pluck it took our breath away. Down he went into the cave without a moment's hesitation, somehow manoeuvring his tender liver around the limestone shelves and projections, to emerge several hours later with the vigour of a cork leaving a champagne bottle. After a spontaneous round of applause he settled down to display his samples, telling us what he had seen and giving us his interpretation of the deposit.

Rhys said the place was a treasure-trove of bones, laid down by ancient streams that had coursed through the caverns. He doubted, however, whether we could accurately date the remains using OSL, a technique that Bert, Rhys and I had hoped to use to date megafaunal extinction. To succeed would be a world first, so Rhys's assessment was disappointing news indeed.

Despite its recent application, luminescence dating (of which OSL is one branch) has venerable origins dating back to the seventeenth century and the Honourable Robert Boyle, founder of the Royal Society and the father of modern chemistry. A bachelor, Boyle lived with his sister Lady Ranelagh at her home in London. 'He is charitable to ingeniose men that are in want,' wrote his contemporary, John Aubrey, 'and foreigne Chymists have had large proofe of his bountie, for he will not spare for cost to gett any rare Secret.' Nothing fascinated Boyle as much as things that glowed, glimmered or shone.

His researches into luminosity were diverse. He made close observation of rotting fish (which sometimes glows feebly) and attempted to import from the Virginia colony a species of flea that reputedly glowed in the dark. But it was his experiments in pursuit of phosphorus that most disturbed the neighbourhood. Boyle knew that phosphorus was derived from 'somewhat that belonged to the body of a man', and in his first series of experiments his unfortunate manservant Bilger was required to collect the pisspots of Pall Mall and boil gallons of urine in the backyard. A disgusting black mass was all that resulted, so Boyle changed tack. Human faeces, he decided, must be the source. Bilger was then ordered to collect and bake huge tubs of the stuff, after which the long-suffering manservant searched for more congenial employment. Boyle did eventually purify phosphorus from human urine (a very poor source of the material) by using greater heat.

It was around 1663 that Boyle made his signal breakthrough in luminescence. He had begun to experiment with an object called the 'virgin carbuncle'—a type of diamond reputed to glow only once, when first heated, and never thereafter. Boyle withdrew to his four-poster bed with the jewel, where he 'brought it to some kind of glimmering light' by 'holding it a good while on a warm part of my naked body'. This excitation of the carbuncle by bodily warmth delighted Boyle, for he was well aware that his constitution 'was not of the hottest'. Over three hundred years later it would also delight Bert, Rhys and me, for it would help date the demise of Australia's giant kangaroos.

In 1948 it was established that the carbuncle's glow came from electrons trapped in the flawed diamond's crystal lattice. As they became excited by the warmth of Boyle's belly button (or wherever he secreted his 'virgin') they escaped their traps and in the process emitted a feeble light. The knowledge that electrons can be trapped in crystal lattices, and then counted when they are released through heating or light, laid the basis for modern thermoluminescence dating.

The technique used by Bert exploits the fact that sand grains are also crystals that trap electrons within their lattices. The energy imparted to the electrons is provided by the Earth's radiation, and once a quartz grain is buried, the 'electron traps' in its lattice will fill at a rate determined by the intensity of this radiation. If you know the rate of background radiation at a site (and thus how quickly the electron traps will fill), you can measure the length of time the sand grain has been buried.

The only problem I foresaw with our application of this ingenious technique was that it was based on dating grains of sand, not bones. How could we be sure that the bones we wished to date were the same age as the sand that enclosed them? This was not a trivial problem, for bones buried in swamps, riverbanks and sand dunes can be re-exposed by drought or flood (which is how we usually find them), then reburied in younger sediment, perhaps becoming mixed with younger bones and artifacts in the process. Exactly the same thing can occur in caves, where underground streams or slumping of floors allows the reburial of bones in younger sediments. Faced with several instances where it was conclusively proved that megafaunal bones had been interred in younger sediment (including the site I had excavated at Minhamite), we decided that the only safe way to proceed was to trust dates only where the sand was found encasing an articulated skeleton or part thereof.

Our reasoning was that it is virtually impossible for a stream or flood to move an articulated set of bones from one deposit to another without disturbing them. While such a conservative approach meant discounting many sites where disarticulated remains were buried in their original context, it had the overwhelming virtue of not including sites where bones had been reburied.

One key problem was locating enough sites. Bert, Rhys and I were surprised at how few ice-age sites there were in Australia that had articulated megafaunal remains. We were unsure whether articulated bones were in Tight Entrance Cave, which is why Rhys so bravely entered it to discover that, with a single possible exception, the bones were disarticulated. We could date the sand in Tight Entrance Cave, he concluded, but a question mark would hang over the age of the bones themselves.

Fortunately, Western Australia's Margaret River region is riddled with caves, and other caverns did contain the articulated remains of megafauna. The most important is Kudjal Yolgah, which lies nestled amid a majestic grove of karri trees just south of Tight Entrance Cave. Thankfully, it is a walk-in cave, and it contains the remains of a modest diversity of large marsupials, including an extinct wombat, a large form of the western grey kangaroo and two species of short-faced kangaroos. After documenting several such deposits, and gaining material from them suitable for dating, we turned to eastern Australia.

The Liverpool Plains of New South Wales is famed for its ice-age fossil deposits, and after returning to Sydney I received news of a discovery in this region by the owners of a property enticingly called The Mystery. It was not easy to find, but eventually I located an old wooden farmhouse with return verandahs, tucked away at the end of an ill-marked dirt road. The man who had found the bones introduced himself as John. He was in his fifties, and he worked the property with his brother and mother. Neither boy had married, and the house seemed to have changed little since their childhood, except for the collection of fossils that stood proudly on the verandah, right beside the front door. All of the familiar megafauna were there—teeth and jaws of giant wallabies, marsupial lions, and the limb-bones of the mighty diprotodon—nine species in all. John had found the fossils on the banks of the Mooki River, which flowed behind the house.

The collection was the work of years, with most finds made after floods had scoured the area, revealing fresh fossil beds. Although none of the bones appeared to be articulated I hoped that such examples might exist at the site itself. As soon as we reached the riverbank and were shown where various finds had been made, however, my disappointment grew, for the bones were from a coarse gravel that formed the banks of a river severely degraded by grazing and salination. There was little chance that any skeletons would be found in such deposits, for the ancient river that carried the gravels into place would likely break them up.

Incidentally, if you are struck by the number of eroded pastoral properties I have mentioned here, it is no coincidence. Erosion is the palaeontologist's friend, and many is the fine deposit that I have seen disappear under the carpet of green spread by Landcare programs. I'm a great supporter of Landcare, so it leaves me feeling distinctly odd, this conflict between good land management and the pursuit of knowledge of the past. I do sometimes find myself praying for the odd monster flood, to gouge at riverbanks and wash away vegetation, so that I might find some interesting fossils.

It was with little enthusiasm that I took a dating sample. This is done by hammering a PVC tube into the fossil-bearing sediment, removing it, then wrapping it in black plastic to protect the sand grains from exposure to sunlight. The dosimeter—a phallic-shaped object—must then be inserted into the hole the sample came from to establish the 'dose rate' of background radiation. After repeating the procedure nearby I packed up and lugged the bulky equipment to the truck. It was a process I would repeat again and again in wide-flung areas of Australia and Papua New Guinea as we pursued our three-year-long study. The Mooki River gravels, incidentally, proved to be between 38,000 and 46,000 years old, but how much older, if any, the fossil bones were was anyone's guess.

Bert and I realised that unless we developed a more economical method of locating suitable fossils we would never complete our work on time. Perhaps one megafaunal site in ten would yield articulated bones, meaning visiting hundreds of sites before finding twenty we could use. There had to be a short cut. Perhaps, we reasoned, if museum curators had neglected to clean the bones of megafauna in their collections, we could obtain enough sediment samples straight out of museum drawers? Museums thereafter became our major research locations, and in them we found much unwashed treasure.

The Queensland Museum yielded a cluster of articulated skeletons found a few years earlier on the Darling Downs and thankfully samples of the sediment that encased them had also been collected. And a series of skeletons with enclosing earth still adhering was also identified in the collections of the Museum of Victoria. In all we obtained samples from twenty-nine sites, but only eighteen had articulated remains.

Then came a wait of months, while tight-lipped Bert conducted his painstaking analysis. He revealed his preliminary results to no one, not even his co-workers. By the end of 2000, as the silence persisted, I was becoming fearful that our work would come to nought. After all, although we picked the most recent-looking sites there was nothing to indicate that they did not span half a million years or more. If the dates were evenly spread in time, the chance of us finding even one site from the last few millennia before megafaunal extinction was remote.

Then I received a phone call. Bert was ready to reveal his findings. This was a special day, he explained, one to be relished, for today there would be just two people on Earth—he and I—who knew when Australia's megafauna became extinct. Kudjal Yolgah Cave in Western Australia and the site from Queensland's Darling Downs, he said, both dated to around 46,000 years ago, give or take a few thousand years (all dates discussed here have an uncertainty of a few thousand years). No site with articulated skeletons dated to after this time, but a large number dated to just before it. The result could hardly have been better, for the two sites are located on opposite sides of the continent, and between them contained a moderate diversity of megafaunal species (rather than one or two species as many sites did). Taken as a whole, our data indicated that a simultaneous extinction event had affected a significant portion of Australia's megafauna around 46,000 years ago.

As our studies wound down and we prepared to publish our findings in the journal *Science*, Bert and other colleagues were using OSL to further refine the time of arrival of people into Australia.

Devil's Lair is one of the most important human occupation sites in Australia. Located a few kilometres north of Kudjal Yolgah, its long, uninterrupted sequence of sediments document climatic, faunal and cultural change in the region for over 63,000 years. Scientists used a battery of dating techniques, including OSL, to date the initial occupation of the shelter to around 46,000 to 48,000 years ago. The equally significant site of Australia's oldest human burial—Lake Mungo in western New South Wales—was also redated using a variety of techniques. Previously thought to be 60,000 years old, initial human occupation there was shown to be around 45,000 to 47,000 years ago. This coincidence with the time of megafaunal extinction was concordant with my hypothesis, but the story is not finished yet. A site in Arnhem Land has a scatter of stone tools in a layer of sand dated to around 53,000 to 60,000 years ago, and a second site may be equally old. Did humans reach and settle Arnhem Land 10,000 years before spreading south? Or did the stone tools drop from higher levels into the older sediment through cracks or other disturbance? At present we simply do not know.

At the time these exciting new dates were coming in, Chris Turney, a bright young researcher from Queen's University, Belfast, was applying the refined ABOX method of carbon-14 dating to vegetation change in Australia. This technique involves rigorous pretreatment of the carbon samples to remove all contamination, then analysis in a mass spectrometer, enabling accurate dates of up to 60,000 years to be obtained. The focus of his research was to date the sediments that record the dramatic alteration in Australia's vegetation, from fire-sensitive to fire-promoting species. The critical sites are in Queensland—with the most important being a crater lake on the Atherton Tablelands known as Lynch's Crater. ABOX dating revealed that the change in vegetation there occurred around 45,000 years ago, again coinciding with our dates for megafaunal extinction and human arrival.

Science is all about testing and re-testing hypotheses, and Rod Wells of Flinders University has been rigorously testing the *Future Eaters* hypothesis by analysing a series of sites in South Australia. He is trying to determine whether megafaunal extinction occurred earlier in the inland than on the coast, as might be expected if climate were the cause. So far all mainland sites with articulated remains he has tested are more than 46,000 years old, with little difference in age between inland and coastal sites. As part of this project an extensive and continuous record of fauna from a cavern called Wet Cave in the Naracoorte area has now been dated. The sedimentary sequence, which lacks megafauna but contains abundant remains of grey kangaroos and other large surviving marsupials, extends back 45,000 years, establishing that megafauna has been absent from the area since that time. More intriguing discoveries were made by Rod on Kangaroo Island, where articulated skeletons of several species of megafauna have been unearthed. With preliminary information suggesting that they may be less than 46,000 years old, exciting work is ongoing at this site.

A second group of researchers led by Judith Field and Richard

Fullagar is examining fossils at Cuddie Springs in western New South Wales. They believe this site indicates that the megafauna survived until around 35,000 years ago, and that it overlapped with people for at least 6000 years. But the skeletons at Cuddie are not articulated, so OSL cannot be used to date the bones. The technique has revealed, however, that the sand grains from the crucial layer in the deposit are of different ages. Perhaps a mass flow of sediment, which did not expose all the grains to sunlight, caused this. Perhaps that flood also jumbled bones and artifacts of differing antiquity in the layer, for the tooth of an ancient crocodile has been found out of place there. It might also explain the co-occurrence of grindstones (of a kind used for crushing grass seed) that elsewhere in Australia date to less than a few thousand years old, with the bones of diprotodons.

Importantly, the advances of the last decade have removed one hypothesis from the competition; that a great aridity at the peak of the ice age 25,000–15,000 years ago destroyed the megafauna. This must be incorrect, because all evidence indicates that the megafauna had vanished continent-wide some 20,000 years earlier, at a time when Australia's climate was a just a little drier and colder than at present. We have no evidence of any major climatic anomalies occurring at this time, although the record is not finely dated enough to have registered a short, severe event, such as a drought lasting a century or less.

It is increasingly evident to me that the magic millennium—the one in which modern Australia was forged—is the forty-sixth before our own. This, it seems, is when the megafauna become extinct, people arrive in Australia and the fire genie is let out of the bottle.

All of this bodes well for the *Future Eaters* hypothesis, but I must emphasise that scientific hypotheses can never be proved, only disproved; and it was with this in mind that Bert Roberts, Chris Turney and I cast around for a simple, elegant and independent experiment that might do just that.

Throughout its history Tasmania has been intermittently joined to the mainland, allowing plants, animals and humans to enter from the north before being cut off by a rising sea. The earliest evidence of humans in Tasmania dates to around 35,000 years ago. Was it possible, we wondered, that an ancestral Bass Strait had kept people out of Tasmania until that time? If so, we had a perfect test for the hypothesis: if we could demonstrate that the Tasmanian megafauna became extinct 46,000 years ago, while people arrived 10,000 years later, then humans could have had nothing to do with the extinction. If, however, Tasmania's megafauna survived until the arrival of people 35,000 years ago, then climate could be ruled out as the causative factor; this is because it is difficult to imagine a scenario where Victoria's megafauna, a few hundred kilometres to the north, was exterminated by climate change 46,000 years ago while leaving Tasmania's giant marsupials abundant and widespread.

Only a new dating program could provide the answers, and it was in search of preliminary information—sufficient to support a grant application—that in August 2003 Chris Turney and I stood in an old Hobart storehouse watching a forklift shift pallets full of rocks, bones and artifacts. We were looking for a set of cardboard boxes that held thousands of bones of Tasmania's megafauna, excavated from a swag of sites right across the Apple Isle. As we opened box after box, we discovered to our joy that few of the bones had been cleaned. Though barely studied, they indicated that Tasmania was home to a curious, insular megafauna that included a small marsupial lion and several distinctive short-faced kangaroos. As I write, our Tasmanian samples are being subjected to a series of tests.

Despite the progress of the last decade, the last word is yet to be had in the great megafaunal extinction debate. And, as always, we may just be a single bone away from a revolution in our understanding.

19

World Conquest

After the extinction of the megafauna, the kangaroo's grip on Australia strengthened. Now more than nine out of every ten species of surviving ground-dwelling marsupial herbivores belonged to the kangaroo family—fifty-one species in Australia and twenty more in New Guinea. It was a triumph that had yet to reach its apogee, for around 8500 years ago wallabies began turning up well beyond their natal shores, indicating that the family had begun to go global.

Fifteen thousand years ago the sea was over 100 metres lower than it is today, allowing Australia, New Guinea, Tasmania and the islands of their continental shelves to coalesce into a single landmass of over 10 million square kilometres. Known as Meganesia it was—with a single exception—the only place on Earth to see a kangaroo. That exception is the spire-like Goodenough Island off southeastern New Guinea. Surrounded by deep ocean, it is a continental fragment that has been isolated for millions of years, and the forests ringing its summit are home to a distinctive forest wallaby known as the black dorcopsis

(*Dorcopsis atrata*). If you ruffle its black pelt an underfur of startling white is revealed, and often one or both paws of this collie-sized creature is also white. In the dense and mossy tangles it inhabits, where sunlight is seldom seen, these may serve as signals. The black dorcopsis is a relic from an earlier time, and may have reached its island home overland millions of years ago. Little changed in its island home, the dorcopsis gives us some idea of what the forest wallabies of an earlier age were like.

Around 8500 years ago, other dorcopsises began appearing on islands that had never been part of Meganesia. This story was revealed as archaeologists probed caves throughout the southwest Pacific as part of a research initiative known as the Lapita Project, which was aimed at exploring the expansion of the Austronesian people. They are the ancestors of the Polynesians and several other peoples, who today are spread from Madagascar to Easter Island. The project took its name from an Austronesian style of pottery known as Lapita ware, and it resulted in a detailed chronology of the Austronesians, who were, before the European imperial age, the most widely distributed group of humans on Earth. It also revealed an extraordinary story of ecological change in a region that spans two-thirds of the globe.

One of the most astonishing discoveries pertinent to our story was made in January 1991 by Dr Peter Bellwood, an archaeologist with the Australian National University. He was working on the starfish-shaped island of Halmahera in eastern Indonesia in 'Gua Siti Nafisa'—the cave of Miss Nafisah's dreaming. Whatever Miss Nafisah dreamed there, she could not in her wildest imaginings have guessed what lay buried beneath her nodding head. Bellwood sounded the sediments of the cave floor, unearthing bones, stones and shards of pottery that spoke of forgotten millennia of human existence. In the lowest levels he found

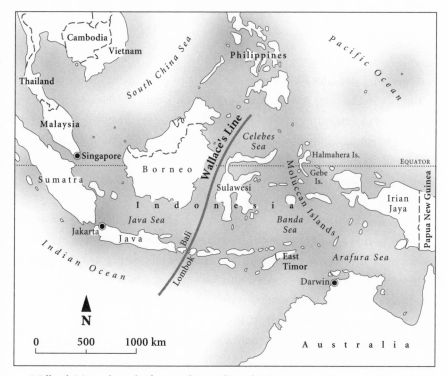

Wallace's Line, where the faunas of Australia and Asia meet.

several jawbones of a dog-sized creature he could not identify. These he brought back to Canberra, and it was with some astonishment that I found myself in Peter's lab later that year, pronouncing them to be the jaws of the chocolate-coloured forest wallaby of southwestern New Guinea, known as *Dorcopsis muelleri*. Such creatures are utterly unknown in the entire Moluccan region (including Halmahera) today. Indeed the cave that harboured the bones lies in the northern hemisphere, near Wallace's Line (where the faunas of Asia and Australasia abut), separated by hundreds of kilometres of open sea from the wallaby's ancestral Meganesian homeland.

A full study of the bones revealed that forest wallabies thrived on Halmahera for millennia, but between 3000 and 1900 years ago they vanished. Peter later found abundant remains of the same creature, this time dating to 8500 years ago, in caves on the island of Gebe, which lies between Halmahera and New Guinea. In both cases the wallabies vanished when the ancestors of the Polynesians settled the islands, bringing with them pottery, new plant crops and dogs. These people also began to replace the original Halmaherans, a process that continues to the present.

How did the wallabies reach the islands? They were, it seems, among the very first animals to be deliberately introduced to a new home by humans. And it was not the mobile Austronesians that did this, but the Aboriginal people of Gebe and Halmahera. These people almost certainly had only rudimentary watercraft—perhaps dugouts or rafts— for the movements occurred long before the great double-hulled Austronesian sailing canoes appeared. For the introduction to be successful several animals would have had to be carried on a long journey. Such long-distance voyaging so early in the Australasian region is itself astonishing, but anyone who has struggled with a wild forest wallaby will realise that the creatures on those cramped watercraft were almost certainly tame—perhaps the joeys of females which had been hunted. Studies of the fossil wallaby's teeth even indicated a potential source area—Missool Island—off the southwestern coast of New Guinea.

But why would people carry tame wallabies from one island to another only to let them loose in the jungle? I suspect that their motive was similar to that of the nineteenth-century acclimatisation societies which brought rabbits, foxes and deer to Australia—to enrich the game available on their island homes. Both Gebe and Halmahera have arisen out of the ocean too recently for an extensive fauna either to have reached them or to have evolved there; their indigenous mammals consist only of bats, rats and possums. As far as we know, Halmahera's Aborigines did not fully domesticate the wallabies, yet their taming and transport represent the earliest human manipulation of a kind that would lead to animal domestication and agriculture. Their achievement is thus a signal landmark on the road towards human domination of the globe.

Other studies done under the auspices of the Lapita Project reveal that such transportation was not unique to the Moluccan Islands. Ample evidence has been found that the New Guinea pademelon (*Thylogale browni*) was introduced by Aboriginal people into a swath of islands lying to New Guinea's north and east. The best documentation comes from the large island of New Ireland in the Bismarck Archipelago, where several excavations record its abrupt arrival around 7000 years ago. New Ireland is a recently formed landmass whose only indigenous land mammals were two species of rats. People have also introduced two possum species, a rat and the pig to the island, so the pademelon was hardly alone in being carried across the sea. From New Ireland the wallaby spread as far north as the islands of Tabar and Lihir, and eastwards to Buka Island in the northern Solomons. (On Buka, however, just a single foot-bone has been found, which may have arrived in a traded skin or other artifact rather than as a living animal.) Still, the spread of kangaroos through an intricate archipelago lying north of Australia, and extending eastwards for over 3500 kilometres, is an astonishing expansion. This, however, was only the first stage in the kangaroo's world conquest.

Since 1791 Europeans have carried kangaroos across the globe, the first living one arriving in London that year. A century and a half later, wild populations had become established in Hawaii, New Zealand and even Europe itself. The Hawaiian population originated around 1916 from a single pair of brush-tailed rock-wallabies (*Petrogale penicillata*) that escaped from a menagerie on the island of Oahu. Today their descendants inhabit rocky slopes in the outer suburbs of Honolulu, and are diverging so rapidly in genetic structure and appearance (having become redder and smaller) that scientists have predicted the Oahu population will eventually be proclaimed a separate species.

New Zealand is home to six different kinds of kangaroos, five of which were released on Kawau Island in 1845 by Governor George Grey, who was governor of South Australia before being posted to New Zealand. He brought with him parma wallabies (*Notamacropus parma*), tammar wallabies (*Notamacropus eugenii*), brush-tailed rock-wallabies, black-striped wallabies (*Notamacropus dorsalis*) and swamp wallabies (*Wallabia bicolor*). Grey performed a valuable service to conservation in bringing the tammars to Kawau, for within eighty years of their trip across the Tasman the population that they had been drawn from (mainland South Australia) was extinct. Now, under the auspices of a New Zealand-born premier of South Australia, New Zealand tammars have been returned to restock national parks where they have been absent for nearly a century.

But it is the red-necked wallaby (*Notamacropus rufogriseus*) that has proved to be the most remarkable traveller. The Tasmanian red-necked wallaby is a hardy creature used to foraging in alpine snows—a capacity that has assisted it in becoming established as far afield as New Zealand's North Island, Britain's Peak District and Ashdown Forest, and Germany's Black Forest (where it thrives on the estate of Prince Reuss). World War II boosted the European populations of this robust Tasmanian, destroying fences around zoos and menageries and allowing it to hop to freedom through bomb-holes. The South Island of New Zealand alone was home to 750,000 before control programs in the 1960s reduced their population to around 3500.

A colony of grey kangaroos, which now numbers around fifty

individuals was established more recently in the Rambouillet forest west of Paris. It originated thirty years ago in a bungled theft of animals from the Emance nature reserve. The locals now seem quite proud of their kangaroos—the Emance school magazine is called the 'Joking Kangaroo', and the town's mayor says that the roos are just part of the local scene.

With colonies of kangaroos now firmly established in Europe and the Pacific Islands, it will be interesting to see where kangaroos next appear in the twenty-first century.

A Dingo-driven Revolution

A profound revolution occurred throughout Australasia over the past 5000 years, the key elements of which were encapsulated by Rhys Jones and linguist Nicholas Evans in 1997:

> ...dramatic changes took place in the human population of Australia some five millennia ago. In the archaeological record this shows up as the emergence of the 'small tool tradition' using flaked points, hafted adzes or microliths, but also as evidence of large-scale gatherings, advances in plant food technologies (especially cycads and seed grinding), the arrival of the dingo, and exploitation of more marginal environments.

Archaeologists refer to this phase of cultural development in Australia as one of 'intensification'—more archaeological deposits, more cultural diversity and more people. It is a change as profound in its own way as the arrival of agriculture in Europe. But what could have precipitated such a revolution? It was, I believe, launched by the arrival of man's best friend, the dog. Let us begin by asking why Halmahera's wallabies vanished. The island is a quarter the size of Tasmania, yet so rugged that even now parts of it remain uninhabited making extinctions by human hands alone unlikely. At around the same time Gebe lost its wallabies, and two species of pademelon (*Thylogale*) vanished from the rugged alpine grasslands of West Papua (previously Irian Jaya). These extinctions were, I believe, caused by the dog, which had been carried to Australia around 4000 years ago by Austronesians. (It may have arrived later in New Guinea and Halmahera.) But why would dogs, which spread throughout Australasia, cause wallaby extinctions only on Halmahera, Gebe, and in the highlands of West Papua? Despite a moderately good fossil record, fossils of thylacines have never been recorded from any of these regions. In places such as Papua New Guinea and Australia thylacine remains occur where similar wallabies survived. Perhaps being hunted by thylacines taught wallabies how to avoid being eaten by dogs.

The dingo has been implicated in the extinction from the mainland of the thylacine, Tasmanian devil and Tasmanian native hen, but evidence for a far more surprising impact comes from studies of Aboriginal languages. One of the most perplexing mysteries of Australia's past is how Aborigines inhabiting over seven-eighths of the continent came to speak dialects derived from a single, recent language family. Known as Pama-Nyungan (Pama and Nyunga mean 'person' in two geographically distant Australian languages), its dialects are as differentiated as the Indo-European languages. Only in parts of Arnhem Land and the Kimberley do more ancient languages survive.

In 1997 Rhys Jones and Nicholas Evans pinpointed the time and place of origin of this successful language family. It arose, they demonstrated, around 5000 years ago (around the time the dingo arrived) in eastern—quite possibly northeastern—Arnhem Land. This is striking information because today the putative ancestral home of Pama-Nyungan is surrounded by people who speak utterly different, more ancient languages. So how was Pama-Nyungan so successful at displacing all other languages to the south of it, but not its immediate neighbours? In this it is resembles English, which has replaced the native languages of North America, but has been unable to drive Gaelic from the British Isles.

Often where languages have spread rapidly, technological innovation has been cited as a cause. In the case of Britain it was (in part at least) maritime innovations and the industrial revolution. The Indo-European language group, which some 7000 years ago began to spread from a source in western Asia into Europe and India, provides another well-studied example. The domestication of the horse has often been cited as the primary cause, though agriculture is becoming a more favoured explanation. For Pama-Nyungan, both the coincidence in timing, and the pattern of its spread, indicates that the acquisition of dogs was a likely cause. Pama-Nyungan speakers would have had little advantage over their neighbours whom, one assumes, had also acquired dogs from the Austronesians on the northern coast. But the dog-less people to the south may well have found themselves at a disadvantage.

At first it is difficult to see what advantages dogs could have brought to Aborigines, for in historic times they were primarily used for warmth and companionship and, except in dense forests, they were not valued as an aid in hunting. Indeed, noisy dogs can be a disadvantage for hunters on the open plains, and Aborigines often go to considerable lengths to leave their dogs in camp when they hunt. It is important, however, to remember that this situation is the outcome of 4000 years of accommodation and learning about each other by dogs, people and marsupials. A quite different situation may have existed when

Pama-Nyungan first began its spread. Then dogs were a new element in the Australian fauna, and the marsupials had yet to learn how best to evade them.

So let us imagine Australia 4000 years ago. Some indication of what it was like can be gained from the historic situation in Tasmania (which dingoes never reached), and by looking at the impact of dogs today. From their arrival in 1803, Tasmanian settlers noted the abundance of marsupials, and in distinct contrast to the early mainland settlements, they gained much sustenance from wallabies, possums and kangaroos. Tasmania continues to be famed for its abundant wildlife, and marsupials are still part of the local cuisine.

And what of the scarcity of kangaroos on the dingo's side of the dog fence? Surely this is a fine demonstration of the ability of dingoes to affect kangaroo numbers. Further clues can be found in remembering the naivety of island-dwelling bettongs which have remained innocent of dogs because of geographical isolation from 10,000 years ago, 6000 years before the dingo arrived.

Such evidence suggests that many marsupials were far more prevalent in Australia 4000 years ago. After 42,000 years of human hunting, a balance had been struck between the marsupials and human predators. Then man's best friend arrives. With its exquisite nose the dog can locate wallabies, bandicoots and kangaroos effortlessly and, although the thylacine may have preconditioned them to the presence of a dog-like carnivore, the smaller marsupials had not yet learned to fear the scent and sight of a dingo. So they sit still and are killed, and the meat they provide is shared with the dog's adopted human family. Certain types of game, such as the Tasmanian native hen, were driven to extinction before they could learn to flee, but many others were simply made more scarce, just as foxes and perhaps cats have made many species rarer in historic times.

But what effect could this process have had on people? Did over-harvesting the marsupials provide the Pama-Nyungan speakers and their dogs with an advantage? Imagine the boost given to a clan that could harvest meat twice as rapidly as its neighbours. They would dominate and spread, especially if, at the same time, the dogs were depriving neighbouring people of food by devastating the standing stock of marsupials. We do not know if it was the people or only their languages that spread, yet the vanished languages of southern Australia appear to have been as much victims of the arrival of the dog as were the thylacine and Tasmanian devil.

So what of the factors mentioned by Jones and Evans—changes in tool technology, plant use, and the increased size of gatherings for ceremonies and other purposes? A starting point is to look at the changes that occurred earliest—around the time the dingo arrived. The most noticeable change in the archaeological record at this time is the spread of the 'small tool tradition'. This appears in northern Australia first, before spreading south. Interestingly, a similar development in tool tradition occured in southeast Asia at around the same time, so perhaps these tools represent another Austronesian-conveyed item which spread. There is no evidence that these tools offered any practical benefit; their spread may simply have been a matter of fashion. Aborigines also began to consume cycad fruit, which are toxic and require an extensive pre-treatment, a practice also known in parts of Africa, India, Guam and the Ryukyu Islands. Cycad fruit treatment is thus also a candidate for transfer by the Austronesians, but in this case the advantage is clear—in historic times cycad fruit fed large gatherings, enabling extended ceremonies to occur.

Two other changes came much later. In a comprehensive study of the archaeological record, Monash University archaeologist Bruno David has shown that by 1000 years ago Aboriginal people had begun to use large, flat grindstones to process grass seed. This innovation occurs around the time that trade increases. One trade item that is readily preserved in the archaeological record is large stone points. These were used to tip spears and daggers, and suitable stone for them was quarried at only a few locations in northern Australia. Again, they only appear during the last thousand years or so.

So how, after an occupation stretching back 46,000 years, are we to account for this sudden propensity by Aboriginal people to eat grass seeds and to trade? I feel that the dingo was, once again, the cause. Wherever herbivore numbers are great, grasses forgo sexual reproduction because their sexual organs are nibbled off before they can ripen. Instead they spread by underground runners. With kangaroo and wallaby populations suppressed by dingoes, perhaps Australia's grasses were flowering and setting seed more successfully than ever before, making it worthwhile to harvest the seeds and grind them between flat stones.

In areas where there are no cycads, seed-cakes played a vital role in feeding participants at large ceremonies. They also made travel (and thus trade) easier, for they are durable and can be carried on long journeys. At base lay a shift in the relationship between the sexes, for grinding seed was women's work, yet the product of their labour was a storable foodstuff that could be appropriated by men to provision the great ceremonies so central to Aboriginal life.

In this dingo-driven revolution we see a profound restructuring of Australia's ecosystems and human cultures, which involved a further diminution of the role of large herbivores, and an increase in human population fuelled by harvesting newly available plant foods. This was a dramatic departure from what had gone before.

The Age of Mammals in Australia

Australia's prehistory is as much about 'black holes' as it is about concrete knowledge. So vast are the gaps in the record that whole mammal families may have come and gone without our knowing. It's like having a jigsaw when only one piece in ten is preserved. But in trying to understand Australia's past, we must work with what we've got. Our current knowledge allows five basic stages to be discerned in the evolution of Australia's mammals. These are:

1) The age of dinosaurs (between 100 and 115 million years ago), when Australia was inhabited by mouse- to rat-sized placental mammals and mouse- to cat-sized monotremes, but no marsupials.

2) The Murgon age (somewhere between 65 and 54 million years ago), when the last of Australia's ancient, terrestrial placentals lived, and when the first (mostly rat-sized) marsupials had arrived from South America.

3) The age of the koala beasts (somewhere between 40 and 20 million years ago), represented by the Lake Pinpa fossils, when large

ground-dwelling koala-like marsupials, and koalas, dominated the herbivore niche.

4) The age of diprotodontids (about 20 and 5 million years ago), as documented at Riversleigh and in central Australia, when wombat-like species were the dominant larger herbivores.

5) The age of kangaroos (5 million years ago to the present) when kangaroos proliferated into a variety of ecological niches.

One hundred and fifteen million years ago Australia was still part of the supercontinent Gondwana, and its flora and fauna were more cosmopolitan than today. The bones of dinosaurs are roughly a hundredfold more abundant in these deposits than are mammals, with only two basic types—monotremes and placentals—represented in Australia.

The monotremes are known from the jawbones of three kinds of animals—different enough to represent three separate families. One of these was a platypus-like species, a second was equally large with teeth superficially like a sea otter, while the third was a very primitive type about the size of a shrew. The larger monotremes were adapted to life in water, most likely rivers and lakes, and this humble triumvirate is the greatest diversity that monotremes ever achieved—a golden age attained while dinosaurs still ruled the Earth. The monotremes are an ancient group that with a single exception is known only from Australia. The exception is a platypus whose 63-million-year-old teeth were discovered in Patagonia. It seems to have migrated there from Australia at the very beginning of the age of mammals.

The discovery of placental mammals in Australia was, in contrast, a great surprise—even a shock—to the scientific community. This can best be understood by looking at the position occupied by Australia 100 million years ago. Then, there were two 'supercontinents', both of which would have been around 70 million square kilometres in extent

(Eurasia, the largest modern continent, is by comparison 54 million square kilometres). Laurasia was located in the northern hemisphere and consisted of what would become Asia (minus India, North America and Europe). The other continent, Gondwana, was located in the southern hemisphere and comprised the remaining major landmasses except South America (which for much of the age of the dinosaurs was the most isolated of all continents).

Until 1997 it was thought that the placental mammals had evolved in the northern hemisphere, where the great majority of species live and where the oldest fossils were found. But that year, after decades of fruitless excavation in the farthest corners of Australia, Nicola Barton, a member of Tom Rich's field crew who was working at a place called Flat Rocks, 100 kilometres from Tom's home in Melbourne, unearthed a jaw so small that half a dozen could sit on your thumbnail. On 8 March—the day of the discovery—the field crew threw a party, and without telling Tom of the discovery, invited him down. After placing the minuscule lump of sediment containing the jaw under a microscope, Tom sat stunned amid the merrymakers, repeating to himself, 'Oh my Lord…Oh my God.' His emotions were understandable: the Riches had long ago remortgaged their house to fund their search, and after twenty-three years of living his dictum about having 'the will to fail', Tom had been running on empty.

In the years that have passed since this discovery, over twenty more specimens have been unearthed, making Flat Rocks one of the most productive early Cretaceous mammal sites in the world. No matter that all these scraps would not fill a matchbox—these are the most significant mammal fossils ever recovered from Australia, for they have resulted in the overturning of a venerable scientific hypothesis and a revolution in our world view.

Such was the importance of the first specimen unearthed that its discovery was published in *Science*. The jaw was well preserved, and displayed a number of features that precluded it from being either a monotreme or a marsupial. Its molars were of a very advanced type that can cut, crush and puncture. Such teeth are known as tribosphenic, and only placental mammals and marsupials possess them; indeed they were a vital element of the success of these mammals, allowing them to open up a new 'food frontier' inaccessible to their competitors. The fossils had only three such molars, and a rear premolar that superficially resembled them. Such features are hallmarks of the placental mammals: a bit like our 'kangaroo essentials', they are features that define the placental lineage. Finally, the specimen lacked the 'marsupial angle', a bony strut unique to marsupials. Such a constellation of features indicated that our fossil was an early placental mammal—among the earliest found anywhere.

I spent a year at Harvard soon after this, and there discovered that our claim had been greeted with uproar and disbelief. Ever since the time of Huxley and Darwin, scientists had assumed that the ancestral cradle of the placentals lay in the northern hemisphere. The firmest objection to our identification was not our interpretation of the specimen itself, but to the notion that it had been found in Australia. The refrain I heard all too often was, 'Everyone knows that Australia had no placental mammals except for some recently arrived bats and rats, so how could this fossil be one? It must be something else that just looked like a placental mammal.' Among the more dogmatic denials I occasionally detected base motives. Palaeontologists must compete for funds to excavate, and this sometimes makes acknowledging a major discovery by another scientist difficult.

Even in Australia the discovery proved controversial, for Michael

Archer has long discounted the possibility that the jaws might have belonged to placentals, instead suggesting (though never supporting) the idea that they are monotremes. Yet Archer is not worried by the thought of placental mammals being present in Australia in the distant past, for he has long championed a solitary, 54-million-year-old tooth from a site his team excavated at Murgon as belonging to just such a creature. My own guess is that this tooth, along with the twenty jaws from Flat Rocks, provides evidence for a placental mammal radiation in Australia which lasted for at least 60 million years before dying out by 54 million years ago. The extinction of placental mammals in Australia may be surprising but it is not implausible. Perhaps Australia's infertile soils advantaged creatures with low metabolic requirements, like reptiles and marsupials, over the high metabolism, energy-hungry placentals.

In 2002, two independent studies of the relationships of the major types of placental mammals were published in the journals *Nature* and *Science*. Both analysed very long pieces of nuclear DNA—16,000 base pairs in all—using a technique called sequencing that, while time-consuming and expensive, is the gold standard for establishing evolutionary lineages. Both studies came to the same conclusion, an outcome that is as close to 'revealed truth' as science ever gets. So, what did the studies find?

An astonishing discovery never before even guessed at by anatomists and palaeontologists was that a group of African mammals which seemingly have little in common—including elephants, aardvarks, elephant shrews and the tiny golden moles—had all shared a common ancestor 90 million years ago (and here the 'molecular clock' could be calibrated against an extensive fossil record). This placental lineage, dubbed the Afrotheria, is the earliest split in the placental family tree. South

America's sloths, anteaters and armadillos (a group known as the Xenarthra) proved to be the next oldest lineage, their ancestor becoming isolated (probably in Antarctica rather than South America) around 80 million years ago. All remaining placental mammals, the studies showed, shared a common ancestor less than 80 million years before the present, which inhabited the northern continent of Laurasia.

The unavoidable implication of these studies is that the placental mammals had begun to diversify on Gondwana at least 90 million years ago, perhaps as the continent began to break up. In the light of this finding, the discovery of 115-million-year-old placental mammal fossils in Australia is to be expected. So it really does appear that the theory first postulated over a century ago by Thomas Huxley, that the marsupials are southern hemisphere in origin and placentals northern, needs to be turned on its head; for we now know that the marsupials arose on Laurasia, in the northern hemisphere, and it looks ever more likely that the placentals arose on Gondwana. Despite the palaeontological and molecular evidence that supports it, the Gondwanan origin of placentals is still a hot topic in palaeontology and I suspect it will be some time before the full implications of Tom Rich's discoveries sink in.

A dark age of about 60 million years separates Tom's contentious fossils from those constituting our next age. It is glimpsed through a single locality near the town of Murgon in southeastern Queensland, where clays laid down in a lake that formed in the crater of an ancient volcano have been dated to at least 54 million years ago (remember that this is a minimum date, and that the fossils may be much older). The mammals, which were all shrew- to rat-sized, are still in the minority in this deposit; though now it is not dinosaurs but crocodiles and giant soft-shelled trionychid turtles (extinct in Australia today, though surviving on other continents) whose bones abound. Most were marsupials, and the Murgon fossils likely represent the earliest marsupial immigrants to reach Australia, for they lived when Australia was connected with Antarctica, and possibly South America as well—as the teeth of several resemble South American species.

Alongside these early marsupials lived the last of Australia's land-dwelling placental mammals, only two teeth of which have been unearthed. Their very rarity suggests that the newly arrived marsupials were in the process of displacing them. From then until the arrival of Australia's first rats and mice around 4 million years ago, the continent was free of land-based placentals.

The fossils of one other placental mammal have been recovered from Murgon. The bones of an ancient bat are quite common, indicating that these creatures had arrived in Australia very early in the evolution of the group. Genetic studies suggest that bats originated in the northern hemisphere, so their presence in Australia so soon after their supposed origin is intriguing.

Between Murgon and the next window opening onto Australia's past lies another dark age—this one lasting about 30 million years. This is extremely frustrating, for much occurred during this interlude, including the separation of Australia from Antarctica (45 million years ago), the onset of Australia's drift northwards (at an initial rate of around twelve centimetres per year), and a series of abrupt and profound changes in the world's climate. These events must have had a marked impact on Australia's fauna, and one or more may have led to the evolution of the first kangaroos.

The window following that dark age is provided by the green clays of Lake Pinpa (guesstimated to be 25 million years old, but which may be anywhere between 40 and 20 million years in age), and it opens onto a greatly altered world. The land-based placental mammals have vanished from Australia and the marsupials have diversified into a number of 'giant' lineages. The largest of the Lake Pinpa herbivores were the calf-sized (but wombat-like in shape) ilariids, and the dog- to sheep-sized wynyardiids. The ilariids had 'cuspy' teeth strikingly similar to those of koalas and ringtail possums. Not many modifications are necessary to produce such teeth from those possessed by the marsupials living in Murgon times. Despite their large size the ilariids mark an early phase of marsupial evolution which was superbly suited to feeding upon leaves, which is perhaps why the koala (an ancient lineage that has not changed much since Lake Pinpa times) still thrives in Australia's gum trees.

The wynyardiids, on the other hand, share similarities with wombats and diprotodontids, and may have given rise to both of these groups. As we have seen, kangaroos were present by Lake Pinpa times, indicating that they are an ancient lineage, albeit one that, unlike koalas, has changed markedly. Yet Lake Pinpa gives us only a passing glimpse at an era of great importance.

A gap of uncertain millennia separates Pinpa from our next window into the past, provided by a large number of fossil deposits from central and northern Australia, which are thought to be Miocene in age (around 25 to 5 million years ago). Even though some sediments were laid down in the same geographic region as Lake Pinpa, in them is a very different cast of animals, indicating that a dramatic and possibly abrupt climate shift had occurred. In what appear to be the oldest of these deposits (best represented in the Lake Eyre Basin) a few archaic marsupials such as wynyardiids linger, but alongside them we find primitive diprotodontids (the lineage that would give rise to the largest marsupial of all time), marsupial lions and various possums, some of which are still around today. There are also several kinds of bulungamayines and balbarines.

We can trace an important trend through these abundant Miocene deposits, for in the oldest the creatures are rather small: the diprotodontids sheep-sized at best, the kangaroos rabbit-sized and the marsupial lion no larger than a tabby. Yet as we progress to those laid down nearer our own time, the average size for these lineages increases markedly, until by the Miocene's end the largest diprotodontids, kangaroos and marsupial lions are ten times heavier, or more, than their ancestors. Because the largest species inhabited open habitats, it is possible that a drying climate, which reduced forest cover, was responsible for this trend. Curiously, arboreal species such as koalas and possums hardly change in size at all.

Following the abundant deposits of the Miocene age comes yet another frustrating gap in the record lasting three to five million years. Opening with the 4.46-million-year-old Hamilton site, many fossil localities follow, and this final phase in the evolution of Australia's mammal fauna is dominated by kangaroos.

The fossil record of North America is remarkably complete and here I wish to compare it against Australia's fossil record, determine what role climate change has had in shaping Australia's fauna, and investigate whether our fauna has ever experienced a 'golden age'. So firstly, how does Australia's fossil record compare with that of North America? At the time of the extinction of the dinosaurs three types of mammals existed in North America—placentals, marsupials and a primitive group known as multituberculates. The marsupials dominated both in number of species and abundance, but many fell victim to the asteroid that carried off the dinosaurs. Thereafter the placental mammals—initially only the size of rats—began their rise, so that by around 18 million years ago they were the only mammal group present on the continent.

The Murgon site gives us an insight into a time when Australia's marsupials had only just arrived and were in the process of ousting the competition—a little like the situation in North America 65 million years ago, only here the marsupials were in the ascendancy.

Within a few million years of the extinction of the dinosaurs, some of North America's placentals were already pig- to cow-sized. Likewise, the cumbersome ilariids of Lake Pinpa may be Australia's first experiment in producing pig-sized herbivores. By 55 million years ago in North America the first large, primitive herbivores had given way to the perissodactyls (rhinos, rhino-like creatures and their relatives), and the earliest artiodactyls (camels, sheep, cattle). These two lineages would come to dominate the large herbivore niche in most of the world. They evolved in response to the challenge of eating tough plant food, the perissodactyls being hindgut fermenters (the primitive placental condition) and the artiodactyls foregut fermenters. 'Fermenting' relates to the way plant materials are broken down by bacteria and other organisms.

Hindgut fermenters accommodate their microbes in either a caecum (the same structure, anatomically speaking, as the human appendix) or a colon. The koala's caecum is proportionately longer than in any other mammal, and into it the most nutritious fraction of the food is diverted, the rest being passed as faeces. While serviceable the caecum has a flaw, which leads to ailments in humans and difficulties for marsupials—it is open at only one end. This means that it must be emptied through the same orifice from which it is filled and if a seed obstructs that opening appendicitis can develop. In herbivores the time required to fill and empty the organ dictates the rate at which food can be processed.

Hindgut fermenters that use the colon to host their microbes—such as the wombat and horse—are not so limited. They just keep shovelling the food into the tubular colon and out the other end. It's a solution suited to very large creatures, which can exist on less nutritious yet abundant food-types, which is why perissodactyls (rhinos, horses and tapirs) tend to be large and not too fussy about food.

The artiodactyls (of whom the ruminants form a subsection) are foregut fermenters that process their food in a complex stomach, burping up portions to be re-chewed as 'cud'. The stomach has the great virtue of being open at both ends and located close to the teeth, allowing bits of vegetation that were hastily swallowed to be broken down at leisure. This strategy is suited to creatures that feed selectively on difficult-to-break-down food and need to extract maximum calorific value from each mouthful. Although the artiodactyls arose around the same time as the perissodactyls, they stayed smaller for longer. Foregut fermentation is suited to a diet of grass, which benefits from prolonged mastication and breakdown by stomach microbes. And that is why the farms of the world are filled with pasture rather than forest. The artiodactyls developed one further feature; they have the equivalent of a gear-shift lever in their ankle. The central unit of this device is the 'knuckle' bone used to play knuckles with, and by altering its orientation a sheep, goat or gazelle can change from 'low' to 'high' gear, quickly increasing speed when required.

Australia's hindgut fermenters, like the perissodactyls, flourished in the earlier part of the age of mammals, but are fewer in number today. They fall into two distinct categories—those like the koala that have a caecum, and those that use the colon as a fermentation chamber. The caecum-bearing koala, and possibly its ilariid relatives, flourished very early on, and were replaced by the colon-fermenting diprotodontids and wombats. Present-day wombats have a relict caecum, indicating that their ancestors once had a functional caecum. I think that by moving the site of fermentation from the caecum to the colon, wombats and their relatives gained an advantage over the ilariids.

Clear parallels exist between the kangaroos and the artiodactyls; both are foregut fermenters that remained small compared with hindgut fermenters, yet in the end came to dominate. Furthermore, both developed highly successful and innovative modes of locomotion—the gear-shift ankle in artiodactyls and hopping in the roos. The parallels are not precise, of course, but it is a distinctly similar course of evolution that has played out on the two continents.

How much of the evolution of Australia's mammals can be explained by climate change? In the light of the threat that human-induced climate change is bringing to biodiversity, this is a vital question. Frustratingly, however, our inability to date most of Australia's animal fossil record means that we cannot relate it directly either to known changes in flora or changes in the world's climate. This lamentable situation allows us to draw only general conclusions. Around 50 million years ago rainforests flourished in central Australia, but drying has since restricted them to parts of the east coast. The fossils from Lake Pinpa suggest that the ilariids lived in a well-watered environment, but whether the region was covered in rainforest or not is difficult to say. By the time the Lake Ngapakaldi deposit formed eucalypts were present, but possums and other arboreal creatures that appear to have been rainforest dwellers still flourished in central Australia. It is not until Bullock Creek times, around 10 million years ago, that fauna from an open plains environment appears and what we see here is very odd: instead of a new fauna which developed in response to the open habitat, we find only a few mammal types, such as the cow-sized *Neohelos* and the primitive balbarine kangaroos, which were present in earlier, forested habitats. Perhaps the climate changed so quickly that there was no time for new kinds of mammals to evolve and they were the only ones that could tolerate the new conditions.

By Alcoota times, around 8 million years ago, the Australian fauna was increasingly adapted to drier, more open conditions. Both macropodines and sthenurines (short-faced kangaroos) had appeared, and they thrived in the dry new Australia. It is tempting to think that they originated in dry habitats, but the fossil record contradicts this—the very oldest sthenurine is preserved in some of the last deposits laid down at Riversleigh, intimating a rainforest origin, while the Hamilton site indicates that the macropodines predominated in rainforest environments from an early stage.

Did a drying climate cause a diminution in Australasia's rainforest diversity? The Riversleigh deposits contain a great diversity of mammals, yet it remains difficult to determine how many species co-existed at any one time. The trouble arises from Riversleigh's many separate fossil deposits, which were laid down over millions of years. Because of their small size many have been lumped together in analyses, giving the illusion that species co-existed when in fact they may have been separated by millenia. Despite the drying trend, some modern rainforests remain as diverse as any realistic estimate of Riversleigh's biodiversity. The Upper Fly and Sepik rivers area in New Guinea, for example, is a mountainous region the size of metropolitan Sydney and home to at least 120 mammal (including many marsupial) species—more than one-third as many as recorded for all of Australia. Given the importance of the impact of climate change on Australia's fauna this should be a focus for future research. Unless we can understand how climate affected

Australia in the past, it will be difficult to manage future climate change.

Did Australia's mammals ever experience a 'golden age'—a period during which large mammals achieved great diversity? During the Miocene period (around 24 to 5 million years ago), as North America's climate dried and warmed, the continent's large mammals diversified spectacularly: some camels evolved to become giraffe-like, hornless rhinos evolved into hippo lookalikes, and up to a dozen species of horse co-existed. Today east Africa, with its magnificent megafauna, is clearly in the midst of such an age. The closest thing Australia ever experienced began in Alcoota times when four cow-sized marsupials (diprotodontids and their relatives) co-existed with several species of gigantic, flightless herbivorous birds (dromornithids).

By the ice age, paradoxically, diversity had increased, largely due to the evolutionary radiation of the short-faced kangaroos, which provided the greatest diversity of large mammals Australia had ever seen. I believe that we live at the end of this modest 'golden age', and its decline began with the arrival of humans. Ever since then, to quote Alfred Russel Wallace, we have lived in 'a zoologically impoverished world, from which all the largest, fiercest and strangest forms have recently disappeared'.

The Groote Eylandt

I wanted to learn more about living Aboriginal cultures and their relationship with the land. So when the opportunity arose to undertake a
survey of the mammals of Groote Eylandt in 1989, I seized it. The
Dutchman Jan Carstensz, who saw the place in 1623, named it well, for
Groote is indeed a 'great' island, lying in the western side of the Gulf of
Carpentaria. The waters that surround it are a shade of aqua I've seen
nowhere else, and where it meets the shore the water is fringed with dark
green mangroves and silver beaches, behind which lie red sand dunes
and the maze-like quartzite hills that form the backbone of the island.

To visit Groote you need either to be invited by the traditional Aboriginal owners or by the managers of a manganese mine known as
GEMCO, which operates in the island's northwestern corner. My entree
came when GEMCO commissioned a survey of the animals inhabiting
their mining lease. For once I was not going to wildest New Guinea, so
I decided to take my children—David, six, and Emma, who was
just four. Although liaison with Groote's Aborigines had not been

foreshadowed, I had requested contact. If they were willing to share their traditional knowledge, I believed I stood to learn a great deal.

GEMCO had organised for Murrabudda, an elder from the Alyangula settlement, to spend time with me. When I went to pick him up I found the settlement a sad-looking place—its dilapidated buildings and litter-strewn streets spoke of dispiritedness and decline. Murrabudda, in contrast, a dark-skinned man of medium height in his seventies, was full of life. When the missionaries came to Groote, they persuaded Murrabudda's father to 'put away' one of his wives. The discarded woman was Murrabudda's mother, and without the support of her husband she and her children had difficulty surviving. They had lived off the land, and so Murrabudda knew the resources of the island exceedingly well. For a few weeks we roamed Groote from end to end, Murrabudda explaining where the animals and plants were found and how Aboriginal people used them. He was clearly delighted to visit places he had not seen for years, and seemed anxious to show me as much as possible.

One day when resting in the shade of a quartzite hill, Murrabudda pointed out some rock art in a shallow cave. Along with the fish, croco-diles and other figures representing the coastal environment, there were two distinctly out-of-place animals—an emu and a Tasmanian devil. The depictions perhaps dated to when Groote was joined to the main-land over 10,000 years ago, and they underlined the fact that the fauna of Groote Eylandt is impoverished when compared with that of the mainland, for not only does it lack emus but also the euro and antilopine wallaroo, which are so significant to the people of Arnhem Land. Also absent was Arnhem Land's nabarlek (*Petrogale concinna*), a tiny rock-wallaby unique among land mammals in having a limitless supply of molars that are replaced from the rear as those at the front wear out.

Only two species of the kangaroo family—the agile wallaby and a rock-wallaby—survive on Groote, but they are remarkably common, making them easy to study. Rod Strong, Groote's police sergeant, took me to a place called Makbamanja to show me rock-wallabies. Its virginal beach with white sand and aqua waters looked idyllic, but Rod's warning not to camp near it was reinforced by drag marks in the sand made by a crocodile of considerable girth. We unrolled our swags between rocks fringed with pandanus palms, and watched the stars blazing in the tropical sky. I rose as the first light of dawn stole across the horizon, and crept towards a rocky promontory at the far end of the beach which was already alive with rock-wallabies.

The species found on Groote Eylandt is the short-eared rock-wallaby (*Petrogale brachyotis*)—one of the most beautiful members of the most exquisite genus of kangaroos. Perhaps it is the creature's habitat—that tropical north of sharp colours and clean edges—or perhaps it is my familiarity with the gentle, soft-furred creatures that makes me think so, for they are not the most brightly coloured or largest of rock-wallabies; but on that morning, as they sat grooming their young and warming themselves in the first light of day on rocks overlooking the sea, they appeared to be living a version of paradise.

Because of their preference for steep, rocky ranges, rock-wallabies are patchily distributed, although they can be (or could be) found in suitable habitats throughout Australia. This combination of a broad yet patchy distribution has provided frequent opportunities for isolated populations to evolve into new species, making the rock-wallaby genus one of the largest in the family.

Scientists are still unravelling exactly how many species there are, but since 1970 five new ones have been discovered—most from coastal Queensland—with one striking discovery made in the Kimberley in the

1970s. The creature had escaped detection because it was mistaken for the diminutive nabarlek, but the new species, known for many years as the warabi (*Petrogale burbidgei*), is even smaller. Weighing in at just a kilogram, it is one of the smallest members of the kangaroo family, and with its large eyes, chinchilla-like fur and big bushy tail it is also one of the most endearing.

The warabi was discovered by a survey team that was documenting the mammals of Western Australia's remote north. Someone asked a local Aboriginal man if he knew the animal and was told, 'That's a warabi, boss.' This was taken to be an Aboriginal name until zoologist Ron Strahan pointed out that 'warabi' was most probably a misheard 'wallaby'; though not before 'warabi' had been given as the species name in a compendium of Australian mammals. Embarrassed mammalogists scrambled to rename the creature 'monjon', its proper Aboriginal appellation, by which name it is known today.

Rock-wallabies have redesigned the basic kangaroo body-plan to live in precipitous areas. They have much-reduced nails on their feet and large footpads that at their ends bear fingerprint-like patterns which are used to grip the tiniest irregularities in the rocks. Their tails are remarkably long and thin, and often bear a tuft at the end. Such tails are of little use for pushing the animals along; instead they are magnificent aerial rudders, invaluable for making precise adjustments in body position while in midair. Rock-wallabies are astonishingly nimble: once in a room in the animal facility at Monash University, a pint-sized nabarlek circled effortlessly above me at ceiling level, ricocheting off the brick walls in an endless series of bounds by inserting the 'fingerprints' at the end of its toes in the cracks.

Rock-wallabies were once so common that a thriving trade in pelts existed. As late as the 1860s you could see rock-wallabies scampering over Sydney Harbour's Middle Head. Now, despite their superb adaptations, many species are gravely threatened. Changed fire regimes and the spread of cats and foxes have seen them vanish from most of Victoria, New South Wales, the southwest of Western Australia and the arid centre.

On the drive back to Alyangula, Rod revealed a little of what a hell this seeming paradise can be for the indigenous people of Groote. The initiation of young men, he said, had long ago been replaced by a stint in Darwin's Fanny Bay Jail. And 'Black Christmas' on the island, when the Aborigines received royalty payments from the mine, was utterly demoralising. If a European had a clapped-out car to sell, this was the time to do it—at a handsome profit; the occasion also marked the high point of alcohol-fuelled violent crime.

Rod was feeling particularly low as he told me this, for a day or two earlier he had interrogated a young man who had driven an axe into his fourteen-year-old girlfriend's head. She had been madly in love with him and, very much against her family's wishes, had run away to be with him. He was in his early twenties, and for a few weeks love had blossomed. Then he had gone out drinking, and when he returned the girl asked him why he had got drunk and left her alone.

He was a fine young man, Rod said, gentle and intelligent, but in a fury he had killed her. All the devastated youth said by way of defence was, 'It wasn't me. It was the drink that killed her.' Rod was leaving the island. In all his life as a policeman in indigenous communities across Australia he had never had such a depressing posting.

It was not a kangaroo but a hopping mouse that facilitated a further insight into the difficulties faced by the Groote Islanders. The northern hopping mouse (*Notomys aquilo*) has its headquarters on the island, specifically in sandy areas fringing Groote's east coast. To visit there I needed the permission of Claude Mamarika, the leader of the Aboriginal community of that part of the island. Around the mine, Mamarika had the reputation of being a hard man.

It took a couple of hours to drive to Claude's home, and as we passed the rocky knolls and groves of spindly palms, Murrabudda told me that Claude's mob had moved to this location to escape the influence of the mine and, particularly, of grog. But for all their efforts it had followed them—young men carrying it home along the dusty, potholed road regardless of what anyone said. Just a few weeks earlier, Murrabudda told me, Claude's son had been killed while driving, drunk, along this very stretch.

We pulled up to the shabby fibro cottage in the sand that was Claude's home and I wondered how to ask this grief-stricken elder for permission to fossick around the sand dunes looking for a mouse. The man that answered my knock had a scowl on his face, and for a moment I didn't know what to say; but then Claude's eyes lighted on my son David, and he smiled. He and David talked a bit, and David explained that we had come to look for mice, at which Claude's smile grew even broader, and he pointed to the dunes behind the shack. I left with a deep sense of gratitude to this embattled leader, who was facing not only the disintegration of the traditional life he had known and treasured, but also the tragic death of his son. I'm still dismayed at my country that such heroism, which is required daily of so many of today's Aboriginal leaders, goes unsung and unheralded in the wider community.

The hopping mice proved hard to find, but we finally located several trackways leading to a burrow. Then we dug and dug and dug. We were still following a spiralling burrow a metre deep as the tropical sun stood high in the sky. The burrow widened slightly and I saw movement in the sand. There's only one way to catch a mouse you are digging out— you must pounce on it and hold it firmly. So I pounced on the squirming pile of sand and gouged out a handful. It didn't feel right for a mouse, so I flung the object away from me, and saw a scorpion, longer than my hand and a pasty-white colour, writhing angrily in the sand. Somehow in our digging we had confused the burrows.

The Groote survey continued for several years, and I had the privilege of studying rock-haunting ringtails, ghost bats and northern quolls, and hosting young crocodiles in my bathroom. The latter came courtesy of Charlie Manolis, an expert on reptiles, which brings to mind another Groote experience. I had got up early to check a line of boxtraps that I had set the day before in a large patch of deciduous vine thicket growing behind some dunes. It was a dirty, dusty environment. The ground was carpeted in dead leaves and the tangle of vines and low, prickly bushes forced me to walk bent over. Native rodents love such places, and it was important that I got to the boxtraps before the sun rose too high, stressing any captive creatures. But an early morning start was not my idea of fun that day, for GEMCO had hosted a highly convivial, libation-fuelled barbecue for us the evening before.

Checking such a trap line is normally easy. You look to see which traps have doors snapped shut. If the trap has not been triggered, the ball of peanut butter and oat bait must be tossed out, along with the army of ants and other insects; but the full traps must be picked up and the 'trappee' transferred to a canvas bag for examination. I usually peer into the trap before emptying it, for you can never be sure what you've caught.

On this morning most traps were empty, but ahead, under a bush surrounded by a dense layer of fallen leaves, was a sprung trap. As I bent over to pick it up, a lightning movement caught my eye and I glimpsed, just centimetres from my face, the business end of a king brown snake. Before I had fully registered its deadly presence, I rocketed backwards; the snake moved just as quickly in the opposite direction, slithering away like liquid through the dead leaves that had so well disguised it. I opened the trap and found a terrified mosaic-tailed rat inside—a common native species whose scent had attracted the snake, which had been trying to gain access to its prey when I arrived on the scene.

2 3

The True Experts

Now that I understood something of Groote's fauna, I was anxious to examine the adjacent mainland, and the most convenient place to do that was at Oenpelli, just inside Arnhem Land, near its border with Kakadu National Park. In the middle of the dry season Oenpelli is one of those picture-postcard perfect places. The surface of the expansive lagoon behind the small settlement is covered in waterbirds, while through the heat haze rises a rugged quartzite mountain, fringed with rainforest and riven with massive crevasses and overhangs, so as to resemble a huge, ruined rampart. The massif is known as Injalak, and it contains one of the most astonishing art galleries in the world. From it I hoped to learn about long-term changes in the region's fauna.

On my first morning in Oenpelli, in 1996, the honking of magpie geese and the wistful cries of the whistling kites were incessant. Smoke from dry-season burning hung like mist over the floodplain, seeping through rocks and softening outlines so that the place resembled a work from the 1890s Heidelberg School—a Victorian winter's morning by

Frederick McCubbin perhaps—except that it was already nudging 30 degrees.

At the Injalak Arts and Crafts Centre, overlooking the lagoon, I wait to meet Tony and Isiah, two of Injalak's custodians. There are already a dozen or so people gathered around the fibro building that stores the completed artworks. Outside, half-painted barks and sheets of paper lie interspersed with roasted magpie geese. The scent of wood smoke, goose fat and singed flesh fills the still air as I'm offered a piece. It is delicious—smoky and moist—the same way meat is enjoyed in New Guinea. Tony arrives and we begin to chat about animals. He says that the knob-tailed gecko is a spirit that 'rapes men'. I'm not sure what he means by this, but geckoes are loathed throughout Australasia. In New Guinea people hate their feel, particularly the way the skin of some species comes away at a touch. Perhaps at Oenpelli the great, unblinking eyes of the knob-tails, along with their gory, red-streaked tongues, somehow suggests homosexual rape.

Isiah arrives and the three of us head out to Injalak Hill, where the melodious calls of butcher birds and honeyeaters fill the air as we climb the boulder slope leading to a cliff at the summit. After breaking through a rim of scrubby, deciduous rainforest, we are transported into an astonishing vision. Below a rounded bluff of quartzite is a long rock shelter, and its entire rear wall is covered in the most vibrant art imaginable. Great X-ray illustrations of barramundi overlay kangaroo, catfish, echidna and hand stencils form a riotous montage of images, as lively and unforgettable as the life in the lagoon that inspired it.

We pick our way through the maze of tunnels and paths that lead from one gallery to the next and my amazement grows, for images of mythic ancestors and spirit beings, very different from the styles we have just seen, fill the walls. And it is here that a second aspect of the complex becomes apparent. Scattered about in niches and in crevices under boulders are piles of ochre-daubed human bones, some of which still bear fragments of clothing. I can understand wanting to be put to rest in such a place—so rich in life, heritage and activity.

In one crevasse high above our heads I discover that Injalak is home to a colony of appropriately named tomb bats. They are large for insect-eating bats, and they rest propped up on their elbows, like so many jet fighters awaiting take-off. And there, at my feet, are the droppings and bones of short-eared rock-wallabies. The creatures are less abundant and more wary here than on Groote, but signs of them are everywhere.

At 3 pm it is unbearably hot, so Isiah suggests that we rest at a shaded lookout where we might get a breeze. There, with the whole of the lagoon and Oenpelli laid out below, I make a baffling discovery. Among the bones of wallabies and waterbirds left from dinners past are the last mortal remains of a goat. This mystifies me, for there are no feral goats in Arnhem Land, and no domestic ones nearby. When I show Isiah the bones he bursts into hysterical laughter and points to Oenpelli, saying that long ago it was a mission station, and the missionaries kept goats. There had been quite a kerfuffle when the goats disappeared. Dingoes were blamed, but now I had found his father out. 'Good thing those missionaries didn't study bones and come up here!' he says, cackling with delight.

When we return to the Injalak Arts and Crafts Centre, Isiah introduces me to an older man called Thompson Yulijirri, whose knowledge of animals and Aboriginal lore is unequalled. When we meet he is painting a picture of an antilopine kangaroo (*Macropus antilopinus*), which he knows as 'kolobarr', and a dingo. The creatures are facing each other, almost like animals on a coat of arms, while between are the accoutrements of the corroborree—didjeridu, ochre, boomerangs and spears.

We get on famously, Thompson and I, for he is happy to have someone to talk about animals with. He tells me how the echidna is the pet of the Mimi spirits—those beings that live in the rocks and create lightning and thunder—and how, as his painting suggests, the dingo and kangaroo were once friends. The antilopine kangaroo, the largest marsupial of Australia's Top End is, Thompson says, now very rare around Oenpelli. 'People have to go a long way towards the coast to hunt it,' he says, but it is still well remembered, for when they must travel far the Oenpelli people perform a dance where they ask for the swiftness and endurance of the great roo.

I examine Thompson's painting at some length. The distinctive features of both roo and dingo have been evened out, making the creatures look more similar than in life. Perhaps this is a way of illustrating their ancestral amity; but the kangaroo is painted X-ray style (so that its internal organs and bones are visible) while the dingo is not.

'Don't eat that one, so don't know how he looks inside,' Thompson says of the dingo by way of explanation. I realise that he has painted the genitals of the two male creatures with startling accuracy. The dog, being a placental mammal, has the penis positioned in front of the scrotum, while the kangaroo, like all marsupials, has the scrotum well in front of the penis. This would not be so evident had Thompson not depicted the antilopine kangaroo with a partial erection, the penis protruding from the cloaca. I comment upon this asymmetry in the otherwise symmetrical painting, and Thompson goes silent. It's not prudery that has affected him, for his eyes shift, examining my face with a searching look before darting about to see if anyone is near. With the coast clear he mumbles, 'Secret story, that one.' I don't ask any more questions.

Thompson now seems inclined to paint alone, so I join the others hanging around the shop. They tell me that art can be a trap, with famous artists sometimes hounded by commercial buyers, even on remote outstations. They whisper about cops holding artists in jail until they have finished a painting, and how credit is extended to famous artists needing cash to fly relatives to funerals, with payment in paintings demanded. I also learn some interesting things about traditional Arnhem Land societies. In some areas second daughters were, in traditional times, expected to forgo reproduction to assist with raising their nieces and nephews. They were given 'medicine' and ritual treatment at their first menstruation to destroy their fertility. I've never heard anything like this before, yet it is reminiscent of the reproduction of some Australian birds, where the young of the previous year delay their own breeding to help their parents bring up the new clutch, because Australian conditions are so harsh that the parents can't do the job alone.

My time in Arnhem Land passes too swiftly. I've so many questions that I could stay for months rather than the days scheduled. Fortunately, I have other work soon to take place in central Australia, where the true experts on Australia's marsupials live. And, on a spring day in 1997, I found myself sitting on the banks of Tietken's Birthday Creek, a beautiful coolibah-fringed waterway that drains the Musgrave Ranges in northern South Australia. Around me Aboriginal children are alternately playing and listening as a Tjilpi—an Anangu elder by the name of Ginger Wikilyiri—tells of the creatures he knew as a youth. I have brought a box-full of museum specimens along, and as Ginger gently picks up the stuffed animals, his eyes fill with sadness. He seems to be searching their soft fur for answers as he says that once they were everywhere—then, how quickly, they were all gone.

The plains and ranges hereabout once swarmed with rabbit-sized marsupials, and their disappearance is one of the most mystifying extinction events the continent has ever seen. Twenty-three species were lost, the remainder being confined to offshore islands or remnant patches. By and large, all that is left today are the larger kangaroos and the mouse-sized creatures. But the remarkable thing is that even now there are people like Ginger who can remember eating those extinct animals. Talking to such elders is like consulting an encyclopedia of now vanished desert life—to a biologist this could hardly be more exciting.

Ginger says that he does not know why they disappeared, but a big drought in the 1930s may have had an effect—foxes too. A woman elder by the name of Mungita chimes in, saying that the drought was important. As they pass the museum skins around, they speak at length and with eloquence in their Yankunytjatjara dialect about the mystery. They clearly feel a great fondness for the vanished animals and the way of life they supported, which is undulled by a half-century absence.

I inquire, one by one, about the various species. Once its Aboriginal name is ascertained the information comes out in a precise, almost formal way. This one lived on the sand dunes, that one in the rocks. This one bred in spring, and had two to three young. That one had a single young at a time. This ate leaves and that one insects. Perhaps this is how elders pass on their voluminous, detailed knowledge of the land to the younger generation—as a catalogue full of detail, all of which must be memorised and added to by experience.

So devastated is the mammal fauna of central Australia that even the brushtail possum is gone from Anangu lands. It was once common, Ginger says, and it held on a little longer than the rest. Then Ginger's son Gilbert, who is perhaps in his fifties and has until now been silent, pipes up. 'I know where they went,' he says in a whisper (a tone Desert people adopt when a major point is to be made). 'They all gone down

Ginger Wikilyiri (right), and myself holding a museum specimen of a chuditch.

to Adelaide. I seen them there.' And indeed, although it has vanished from 80 per cent of its habitat, the brushtail possum remains abundant in Australia's cities.

Later that day we return to our camp among the granite domes of the Musgraves. A light rain has fallen, and the lichens on Sentinel Hill now look alive, punctuating the rocks with pale green. Somewhere among those domes are the last of the middle-sized mammals left in the region—three small colonies of black-flanked rock-wallabies (*Petrogale lateralis*). A team from the South Australian Department of Environment and Heritage is there conducting a survey, and has caught a female. I have the honour of holding her while she is weighed and her pouch checked for young. She is a lovely, soft creature with a well-grown joey. Despite her fertility the colonies are vanishing, laid siege to by changed patterns of fire, introduced predators and competition from goats and rabbits. Fox-baiting might give them a chance, the rangers say, but dingoes take the baits too, and dingoes kill foxes, making them important allies. To my dismay I discover that rabbits are—with the aid of a post-calici bounceback—relatively abundant, as are cats.

As I sit among the granite tors discussing the diminishing wallaby colonies, my attention is drawn to an older man who has thus far remained silent. His name is Robin and he speaks very little English. I learn through an interpreter that Robin was born on the other side of the frontier, and has always led an independent life, avoiding the mission stations. His principal early contact with Europeans was through dingo trappers, with whom he travelled on camel and horseback, learning how to take dog scalps for the bounty. Robin is one of only two old men in the area who can still knap stone (strike flakes off larger pieces to make tools), and as someone who lived through a first-contact situation he has interesting things to say about the arrival of the Europeans.

Robin saw his first white man while he was still a youth. Both he and Ginger explain that Aborigines encountering Europeans for the first time stalked the intruders, doing without fire for days so as not to draw attention to themselves. This fascinates me because it bears on an important argument surrounding land management in Australia—how frequent was fire before 1788, when the First Fleet brought its convicts to Sydney Cove? There are those who argue that burning the bush was rather infrequent back then. They explain away the many instances of smoke recorded by explorers as the result of signal fires lit by Aborigines to alert others to the presence of the strangers. But Robin and Ginger's testimony points in the opposite direction: fire may have been suppressed by Aborigines who saw the intruders, so the columns of smoke recorded by Europeans may have been fewer than normal.

Both men also have something to say about the chuditch (*Dasyurus geoffroyi*), a cat-sized, white-spotted carnivorous marsupial that once ranged through the inland, and which they know as 'achilpa'. When they state without hesitation that its main diet in their area was termites, I at first think that these venerable 'professors' must have confused it with the striped numbat (*Myrmecobius fasciatus*), but then I see my western arrogance is getting in the way, for they are sure that achilpa could also kill and eat rabbit-sized creatures.

A few days later, near Uluru, I meet Nugget Dawson (Tjilpi Nagada)—the other capable stone-knapper and a great source of traditional knowledge. As we speak Nugget draws in the sand with his finger, tracing the journeys of his youth in the days before roads, cars, camels and white men. 'Walking everywhere,' our translator says, 'walking all over my country, burning, hunting, visiting the sacred sites, making sure that the red kangaroos increase, the hare wallabies increase.' His finger describes a great oval in the red sand, a trace linking small circles and crinkled lines, representing the all-important waters and sacred sites. Then he stops and looks directly at me, saying, 'But now the white men have come and they have made their own sacred sites, putting up fences around them so that we cannot go in. They use those sacred sites to increase their money.'

He is absolutely right. The whites took the land from the Aborigines and gave it to their sacred cows, in whose name they irrevocably changed the Centre, so that Nugget's remembered landscape is no more, and can never be again.

There is one creature I'm fascinated with, but I did not then know its Aboriginal name, nor could I describe it. Known as the central hare wallaby, its scientific name, *Lagorchestes asomatus*, translates as 'the bodiless dancing hare', which is very appropriate, for except to a handful of people like Nugget Dawson, its body remains a mystery. Europeans first learned of it in 1932 when, in the vast region surrounding Lake Disappointment near the Western Australia–Northern Territory border, explorer and mineral prospector Michael Terry came across what he called a 'spinifex rat'. The skull was given to the South Australian Museum, and there it lay until, a decade later, it was described by the outstanding amateur mammalogist Hedley Herbert Finlayson as belonging to a distinctive member of the kangaroo family. But it has never been seen again, and Terry left us no description of the animal, unless perhaps 'Donald' in the excerpt below belonged to the same species:

> What a pretty little thing Donald was. Shaped just like a kangaroo, the wee thing would curl up nicely on my hand. It had beautiful black eyes, and travelled for hours in my shirt front, warm against my tummy as if in its mother's pouch.
>
> For a day or two [following his capture] Donald sulked; he would not eat the offering of dry grass stalks, would not drink till his tiny nose was pushed into weak milk in his tin. In a week, however, he was as tame as a cat. With a boot lace around his shoulders attached to a cigarette tin holding a few pebbles to act as a drag and advertise his presence, he hopped about the camp. Poor Stan [Terry's companion] got scared to move lest he trod upon my pet which had soon to be tethered clear of harm's way, especially from the attentions of Chou-Chou [Terry's dog]. For that person had natural and definite ideas about Donald; in fact I had an awful moment when the young fellow in all innocence hopped too close to the eager jaws.

Sometime later I learned that researchers from the Conservation Commission of the Northern Territory found that the Aborigines of the Western Desert remember a kind of kangaroo with long, soft grey fur, hairy feet and a short, thick tail which they knew as 'kananpa'. It was, they said, the 'deaf one' or the 'stupid one' because it refused to leave its shelter until it was too late to flee. They report that it survived in the northeast of the Great Sandy Desert until at least 1960; but we Europeans were just too busy tending our sacred sites to go and look for it, so some doubt will always remain over the identity and appearance of that bodiless dancing hare. For my part though, knowing how thoroughly the desert Aborigines catalogued their fauna, I'm convinced that the 'deaf one' was *Lagorchestes asomatus*. Had there been another species, the Tjilpi would surely have known about it.

Symbols of the New Land

The fate of the kangaroos is inextricably bound with the fate of my country, so in tracing the health of their populations we can identify how things stand in this 'fifth part of the world'. So let's examine the kangaroo family as it is today, beginning with the great red and greys. These species are superabundant in some regions, yet are rare or absent in others. Why? To understand this we need to revisit the time when Europeans first took up the land. How were kangaroos faring then?

When reading explorers' diaries and journals it is hard to avoid the conclusion that, prior to pastoral settlement, the larger kangaroos were rarely seen over most of Australia. The naturalist John Gould, who produced a monograph on the kangaroo family and who knew more about kangaroos than any other nineteenth-century European, predicted the extinction of the red kangaroo, so parlous did its situation seem to him. The diaries of the early pastoralists contain much, however, to suggest that within a few decades of European settlement kangaroo numbers dramatically increased. One important account is from Duncan Stewart, who arrived in Mount Gambier, near the South Australian coast, in 1846. Then the settlement was only a few years old. More than six decades later, in 1910, he recalled:

> When the writer came to the district, kangaroos were not by any account plentiful; although some 25 years later, they had become so numerous that the Government and the settlers had to employ men to destroy them, as they were devouring nearly all the feed. They became almost as much a plague as rabbits are at the present time. The dying out of the natives might, to some degree, account for the increase of the marsupials. Some 50,000 were destroyed in five years.

Could Stewart have been correct in attributing the increase in kangaroos to the decline of the Aboriginal population? Biologists have long ignored this possibility, pointing instead to the provision of watering points for stock as the vital factor in the increase. For the less arid-adapted grey kangaroos the provision of water does appear to have permitted their expansion into arid regions, but this cannot explain the entire increase which occurred in the well-watered districts. Changes in pasture quality or a run of good seasons may have helped, but are not enough to explain the continent-wide expansion that invariably followed pastoral settlement.

Stewart was correct, at least in part, in his analysis, but we should not minimise the role played by the dingo in controlling kangaroo numbers. Shepherds were relentless in their pursuit of wild dogs, which quickly vanished from the settled districts; and, as we have seen, dingoes can have a large impact upon kangaroo numbers.

If you have ever been fortunate enough to travel into the Aboriginal lands of central Australia you may have been puzzled by the absence of kangaroos. In 1994, at Utopia station in the Northern Territory—home of fabulous dot paintings and Aboriginal batik—I asked an elder by the name of Quart-pot Corbett about this. Red kangaroos, Quart-pot said, were all but extinct in his country. He put this down to the large numbers of people concentrated in a smallish area, and to hunting with rifles. Over the years Quart-pot had seen young men venture out in their Holdens and Fords with their 303s, unconcerned about traditional hunting restrictions, and bringing back fewer and fewer kangaroos. As I've travelled to Aboriginal communities around Australia it's a story I've heard again and again.

To see large kangaroos it is best to visit the pastoral areas south of the 'dog fence' where the kangaroo shooters operate. You might imagine that the kangaroo industry would have reduced kangaroo numbers in these regions, after all it harvests millions of them every year—in 2003 the figure was 6,500,000. (The same year it employed 4000 people and generated $200 million in income.) Yet there is no decrease in kangaroo numbers, in part because the industry is designed not to do so (the quota is set at a 'sustainable yield' of 10 to 15 per cent of the estimated total population each year), and partly because the hunting is conducted where dingoes are absent and kangaroos consequently exist in great densities.

In national parks and reserves we sometimes see an even greater abundance of large kangaroos. The history of Tidbinbilla, now a popular nature reserve near Canberra, is typical of many pastoral districts in the decades following European settlement in that by 1870 eastern grey kangaroos were reported to be present in 'plague proportions'. A campaign of eradication reversed that initial boom so that by 1964, when the reserve was proclaimed, kangaroos were all but extinct in the area.

After just thirty years of protection, however, they had returned in force, their density reaching a phenomenal 357 animals per square kilometre and their grazing pressure turning the grasslands into a short-cropped lawn. Faced with mass starvation of the kangaroos, in 1995 the authorities elected to cull nearly 1000 individuals. In a nod to our supposed cultural sensitivities, the bodies of these animals were buried rather than utilised. But six years later the kangaroo population had almost returned to its 1995 level, and managers are again grappling with the issue of culling.

There are other national parks where kangaroos have been exterminated without the opportunity to recolonise. Marra Marra National Park, just north of Sydney, is mostly sandstone country unsuitable for eastern grey kangaroos, but there are a few pockets where they would flourish (and presumably did before European settlement) if only the animals could access them. In the absence of kangaroos these areas today grow a luxuriant crop of grass, which in summer dries off to become an ignition point for fires that rage through the region. In 2002 a massive conflagration scorched the entire park, causing great concern about the maintenance of biodiversity.

In instances where kangaroos have been exterminated, and also where they have multiplied to the point of mass starvation, the balance of nature has been lost and Australia's biodiversity has suffered. How can that balance be regained? In parks like Tidbinbilla the introduction of predators could make a difference, but the two ideal options from a biological point of view—dingo and thylacine—are not feasible, for one is extinct while the other would not be tolerated by sheep farmers. Humans therefore remain the only feasible regulator of kangaroo populations, and whether they act by culling or instituting some form of reproductive control (perhaps a technique for the future), it is important—both for humane reasons and for the sake of the environment—that the option be exercised. As for places like Marra Marra, the answer is to reintroduce the roos, and then control their numbers to reduce the incidence of fire and starvation alike.

The situation on Aboriginal lands is more complex, for the vanishing of kangaroos from these regions speaks poignantly of a country and a people who have, at least momentarily, lost their balance. For 46,000 years Aboriginal people hunted kangaroos, guided by a set of cultural beliefs that allowed the two to co-exist. Today those beliefs and the technology used in the hunt have been forever altered. A new accommodation with the land is needed—one that is in tune with both Aboriginal culture and the needs of kangaroos—and this is something that only the Aboriginal people themselves are capable of devising.

An extraordinary study, done in the 1980s by one of the great kangaroo experts, Alan Newsome, revealed just how delicate the balance between kangaroo, country and people is, and how it can be upset in the most unexpected ways. Newsome was studying a population of red kangaroos living on Bert Plain near Alice Springs when he became fascinated by the low fertility of the group, for fewer than half the females were pregnant under conditions that should have seen all of them carrying joeys. After extensive studies of both sexes he discovered that male infertility was to blame, and that an unexpected pattern of infertility existed in the area. Among kangaroos inhabiting the 'floodouts' where creeks disgorge onto the plains and where trees provide shade, approximately one male in three showed impaired fertility. Out on the Mitchell-grass plains (where the creatures abounded), only one male kangaroo in six was fully fertile. Newsome hypothesised that the continuous high temperatures of the shadeless plains had effectively cooked the testicles of the big bucks, inflicting such damage that many would never recover their full potency. In his perplexity as to how the red kangaroos could be so maladapted, Newsome consulted a 'kangaroo man' of the Unmatjera clan, who sang his kangaroo song, which listed the key totemic places where, in traditional times, the creatures abounded. Newsome says of the song that:

The legends relate basically overland journeys of creation, travelling in the daytime and traversed by natural means, [but] there is a gap of about 120 kilometres (from the second to the third most westerly sites known to me), which is very poor habitat for red kangaroos. The legend describes how a great wind bore one of the heroic kangaroo ancestors across that gap. As well, the Unmatjera legend includes an even longer supernatural means of travel, underground, between two sites separated by desert for a distance of about 350 kilometres. The ancestor finally emerged near today's suitable habitat for the red kangaroo.

Because of his great familiarity with the country Newsome was able to map the places mentioned in the Unmatjera song, discovering that

ten of the fourteen sites lie along or near the most favoured habitat of red kangaroos, the major watercourses. Half of them are on flood-outs, and the most famous totemic site is close to the very best kangaroo habitat in all of central Australia, an extensive floodout.

In the traditional red kangaroo habitat, the fertility of the males is maintained by the cooling shade of trees. So why had they forsaken such country for the open plains, where they are most abundant today? Newsome believes that the reds were drawn to the Mitchell-grass plains after the establishment of the cattle industry, when intensive grazing removed the mature, dry stems from the tussocks, forcing the plants to put forth the tender green shoots favoured by red kangaroos. By promoting a change in the timing and distribution of new growth in grass, the grazing of the cattle had, via the agency of a merciless sun, effectively sterilised the male reds en masse. Newsome's study provided a valuable insight on how delicately balanced life is on this continent, and how easy it is to damage its creatures. It also filled me with fear for the future of the large kangaroos in the face of global warming.

If a few populations of larger kangaroos have waxed so mightily that their very prosperity has become a risk to our collective ecological health, this must be balanced against the underprivileged legions of the extinct and vanishing. Seven species of kangaroos—10 per cent of all that existed 100 years ago—are today extinct. A further seven species—30 per cent of Australia's remaining smaller kangaroos—are now so reduced in number and distribution that they no longer play a functional role in Australia's ecosystems.

O o l a c u n t a !

I'm writing these words in 2003, on my way home from London where I've been studying some of Australia's most interesting mammals. It's been a sad pilgrimage, but a necessary one, for the only surviving examples of the long extinct creatures I'm interested in reside in London's Natural History Museum. Its collections are the richest and oldest in existence—the spoils of an empire—and they include a treasure-trove of Australian mammals housed in a cavernous storeroom filled with tall green cabinets, whose drawers bear the scientific names of the occupants. Pull a drawer out and you will see their stuffed bodies lying row upon row, as neat as soldiers on parade.

It is a strange feature of Australia's historic extinction epidemic that it struck most fiercely at those species that seemed most secure. The native rats and mice which once swarmed over the inland in countless millions suffered a greater depletion than Australia's marsupials and monotremes; while among the marsupials it was those paragons of success, the kangaroos, which lost the most species.

Looking at the taxidermised remains of broad-faced potoroos, nail-tailed wallabies and desert rat-kangaroos, I feel as if Britain has taken the heart of my country. The desert rat-kangaroo, eastern hare wallaby and crescent nailtail once thrived right across the land stretching out 11,000 metres below me, but the plants they browsed now go unclipped by their dainty teeth, while the tribes and predators they fed must make shift without them. Perhaps, I secretly hope, my studies of the ecology of these vanished creatures will assist in regaining that equilibrium, for until we know what we have lost, we cannot make good the damage.

As I write, outside the plane a vermilion line announces the coming of the day. I've seen it often enough from below, but from up here it is a miracle. Galah-grey clouds stretch from horizon to horizon, sculpted by winds into a monochrome rippled beach, through which the rising sun spills a lava of pink—an eerie rose glow from below, piercing the cloud in strange patches.

My jetlagged mind is suddenly thrown back twenty years, to the other side of that ripplefield of cloud. I'm in the loneliest desert on earth, wandering towards camp after a day spent searching for fossils on the shores of a dry salt lake. I move along the crest of a blood-red sand-dune, its summit a maze of dead-looking clumps of cane grass and sandy blow-outs. It's been a long, hot day, and my water ran out hours ago. There's not a sound, not even the wind, to remind me that I share the Earth with another living creature.

My eyes are trained to scan the ground for the tiniest fossil—I usually find the lost earring, the contact lens, the money on the pavement. Now I see miniature black dragon, its body thrown into an S-shape that is half buried in sand at my feet. My tongue rasps against the roof of my mouth as I muse on the tricks that an exhausted and dehydrated brain can play on you. Surely this is nothing but an oddly shaped stone—a mirage of a fossil of a mythical creature that's been awaiting me in this lonely spot since the dreamtime? I pick it up, but immediately drop it. I can't interpret the sensation in my fingertips, then I realise I've been burned. I bend once more to pick up the mysterious object—more cautiously this time—juggling it to dissipate the heat. Incredibly, it *is* a miniature dragon—made of brass and blackened by time, which has lain there all day on the dune crest storing up the heat of the sun. It is one half of an old Chinese belt buckle that, intertwined with a brother, once upheld someone's dignity. Is it possible, I wonder, that an errant Chinaman, bound for the goldfields of Ballarat, perished out here by Lake Eyre, leaving the buckle as the only testimony to his existence?

Then I see the chips of stone and the bleached and broken bones of a desert rat-kangaroo—like the scattered skeleton of a young rabbit in the sand. This animal was last sighted before I was born—indicating that a Wangkangguru hunter once sat here, relishing a delicious meal of oolacunta, as the tribe knew the marsupial.

Perhaps the lucky Wangkangguru had been given the belt, and maybe a pair of pants as well, in exchange for leading an explorer to water or in payment for mustering cattle for some forgotten pioneer. Or perhaps the solitary half-buckle had been traded in from the coast by Aborigines before the European pastoralists ever arrived, its lively depiction of the dragon and metallic lustre endowing it with a power that moved it along the great trade routes in exchange for pituri or stone axes, until at last it pierced to the heart of the continent.

Whatever the case, the buckle had reached its resting place in the sand by the time of the Great Depression, for that is when the oolacunta was last seen alive. The diminutive creature had the finest lines of any member of the kangaroo family—all legs, grace and energy—with fore-limbs so tiny that they almost disappeared when, in full flight, it tucked them up close to the body. At under a kilogram and with fur the colour of fine beach sand, it was a mere atom of life in the vastness of the inland, an almost ethereal being whose appearance and disappearance is a profound biological mystery. First described in 1843 by the great English naturalist John Gould, who gave no indication that it was rare, the oolacunta promptly vanished into thin air and was not seen again for nearly ninety years.

Almost everything we know about the creature was learned through the agency of a single man—the one-eyed, one-handed Hedley Herbert Finlayson, a chemistry tutor from Adelaide University who braved the desert on camelback during the height of summer in 1931 to investigate sightings of a tiny rat-kangaroo in the far northeastern corner of South Australia. Finlayson is a true unsung hero whose achievements as an amateur mammalogist, at a time when so many scientists were seeking advancement through study at Oxford and Cambridge, read like high adventure. His journeys, which twice nearly cost him his life, were made at a time when many Australian mammals were disappearing, and they give us a glimpse of many now vanished creatures. To find the desert rat-kangaroo Finlayson had to penetrate the endless plains and sand-dunes of the Lake Eyre Basin. At times all he had to eat was curried oolacunta, for he had great success in his quest.

He found the species living in the vicinity of Cooncherie waterhole, and was flabbergasted to observe that, in this hostile country where the temperature at ground level is often in excess of 50 degrees Celsius, the oolacunta never sought shelter in a burrow, instead making do with a loose nest of sticks constructed on the open plain. Despite their size they were as brave and dogged as the largest red. Finlayson wrote:

We had ridden less than half an hour when there came a shrill excited 'Yuchai' from the horse-boy furthest out, and the chase was on... Tommy came heading back down the line towards the sand-hill, but it was only after much straining of eyes that the oolacunta could be distinguished—a mere speck, thirty or forty yards ahead. At that distance it seemed scarcely to touch the ground; it almost floated ahead in an eerie, effortless way...as it came up to us I galloped alongside to keep it under observation as long as possible. Its speed, for such an atom, was wonderful, and its endurance amazing.

We had considerable difficulty heading it with fresh horses. When we finally got it...it had run us 12 miles; all under such adverse conditions of heat and rough going as to make it almost incredible that so small a frame should be capable of such immense output of energy...

In his private notebook Finlayson recorded, in a terse, yet admiring way how the chase ended. 'Finally he staggered and dropped, and lay gasping...Difficult to imagine anything gamer—only stopped to die.'

For *twelve miles* this tiny creature, weighing just under a kilogram—less than a rabbit or cat—had outpaced one fresh horse after another! What I would give for just a single day with an oolacunta, to observe and learn how this most amazing of kangaroos lived, for it is the very epitome of the toughness needed to survive in Australia. As it was we learned shamefully little about it before it became extinct except that, unusually among kangaroos, the females were larger than the males, suggesting that females were dominant.

In London I glimpsed what were probably elements of its success— a massive nasal cavity which gave the head a unique broadness, a black band of fur below its sandy outer layer, and bare patches on its arms, chest and inner thigh. Its huge nasal region may have cooled the scorching air before it reached its lungs, as well as extracting precious moisture from the exhaling breath. The black band in the fur may have retained warmth on a cold desert day—perhaps when the creature ruffled its coat slightly so as to expose it—and perhaps it licked its bare patches, which would have been shaded as it sat, the evaporation dissipating the unbearable heat of summer. Whatever its many secrets, they added up to a unique animal, one that could lie on the open plain in its flimsy nest all day, enduring the worst of desert conditions, but at the slightest danger rise and take off in a straight line—for twenty kilometres if need be—purchasing its survival with unique speed and endurance. Taken as a whole, the creature's strategy makes sense; its flimsy nest was simply a resting place from which an approaching predator could easily be spied. For such a creature, seeking refuge in a cool burrow where it might have been trapped was a far poorer option.

There is the faintest glimmer of hope, fanned in part by its previous near-century-long disappearance, that we may see the oolacunta once again. In the 1970s workers on the dingo fence in western Queensland reported a kangaroo the size of a football running along and bouncing off the netting, and from time to time similar sightings are reported in the Lake Eyre Basin. As with the thylacine, however, every year that goes by without rediscovery increases the likelihood that the species is truly extinct.

There is a rather sad ending to the story of Finlayson and his oolacunta, for when he published his bestselling book *The Red Centre* in 1935, he was taken to task by Ellis le Geyt Troughton, curator of mammals at the Australian Museum, for collecting such rare creatures. It was an attack motivated in part by envy, but Finlayson's amateur status left him vulnerable to such pronouncements—indeed criticism by professionals seemed to dog him all his life. A particular bitterness came in the 1950s and 60s, when the Australian Mammal Society was founded as an association for 'professional mammalogists'. Despite pleas from some members, Finlayson refused to join, for he had been sensitised to the stigma of 'amateur'. For the rest of his life this fascinating man, who lived to be nearly 100, never married, drove a car, or owned a television or phone. He kept his priceless specimens in his house in North Adelaide and only after his death were all 3000 deposited in a collection in the Northern Territory.

From the air the country traversed by Finlayson looks as if its deep geological history has been written in braille, with each dune and ripple telling of bygone winds and weather systems. The dunes run north–south below me, so Uluru must be off to the northwest. I know that because the winds that once marshalled those countless sand grains were generated by a vast high pressure system that sat almost directly over the great monolith. Then—15,000 years ago—Australia was a Sahara, so dry and windy that the heart of my country was little but shifting sand.

Soon Algebuckna waterhole on the Neales River appears, its narrow waterway a glistening thread of silver in the morning sun, pointing straight at Lake Eyre. I imagine the pelicans slowly shuffling in the dawn light, the cormorants at their perches preening as the sun touches the crowns of the red-gums. Then the vast, silvered, extent of Lake Torrens, the Flinders Ranges forming its stately backdrop. After that, Spencer Gulf, and finally touchdown in Adelaide. I'm fresh from the wonders of Europe and America, but this is the most mysterious and beautiful country on Earth.

Re-making Country

When I wrote *The Future Eaters* I put forward a hypothesis developed by Ken Johnson and Dave Gibson of the Arid Zone Research Institute in Alice Springs, that the extinction of Australia's medium-sized mammals was largely due to a changed fire regime that occurred as Aboriginal people left the land. Much new information about this has come from the last of the desert nomads journeying back to their country—some of whom made first contact with Europeans as late as the 1960s. The first opportunity many of these people had to return to their ancestral lands came courtesy of biologists, who hoped to discover populations of mammals feared to be extinct. The Aborigines, many of whom were from the Great Victoria and northern Tanami deserts, were jubilant at this prospect. Life on the missions had not been kind to them; many suffered a sharp decline in health and a younger generation, alienated from the land, was growing up.

The returning nomads were confident that they would find plentiful game, including many supposedly extinct species, which had fed them right up to the moment they stepped out of the desert a few decades earlier. Upon entering their country, however, they detected something wrong, for the unburned vegetation was old, and at a uniform growth stage. As they lit fires many remarked, 'No one's been caring for the country. It needs to be tidied up.'

But it was only when they began to walk through their land that the full extent of the catastrophe became evident—there were no mammal tracks. The country was empty of game, no longer capable of supporting traditional Aboriginal life. It was a discovery that left many deeply disturbed and depressed.

When the elders were asked what they thought had gone wrong, these experienced land managers gave a prompt and, to European ears at least, extraordinary answer. The extinction of so many animals was, they said, due to their own negligence; they had failed to carry out the increase ceremonies upon which the animals depended, and now the creatures had forsaken the land. Although the biologists put things a little differently, in essence they agreed that the Aborigines' traditional lifestyles had been the lifeblood of the country. As they moved about they had burned the vegetation into a complex mosaic in different stages of growth, which many marsupial species required to survive. In effect, the very act of living on the land—of hunting, gathering and burning— had maintained its diversity.

By the time European biologists were taking a serious interest in our desert marsupials, only a single colony of the rufous hare wallaby, also called the mala (*Lagorchestes hirsutus*), survived on the mainland. It held on in a region that had been managed by Aboriginal people until well into the twentieth century, at Dragon Soak in the northern Tanami Desert. Perhaps it was a matter of luck that this population alone survived; nevertheless, I can't help but bitterly reflect that had action been taken a decade or two earlier, we may have saved half a dozen species from extinction.

Nothing it seems could save that last colony of the mala, for despite the best efforts of biologists to protect it, it was exterminated by fire in November 1991. Fortunately, just before this a few individuals were taken into captivity, and there they bred prolifically. Some were released back into the area around Dragon Soak, but these were swiftly eaten by cats. Such are the changes that have occurred to the Australian environment since this once widespread species vanished, that no success has been enjoyed in establishing any wild-living populations anywhere in the continent. So today mala can only be seen in fenced areas, where they are protected from feral predators and fire. Still, the creatures survive, and I'm determined to see them once again living wild in the desert.

You cannot have properly functioning ecosystems unless you have their necessary parts—the plants and animals of which ecosystems are composed. During my lifetime many of Australia's unique species have been let slip to extinction. A couple have been rescued at the last moment, but with each year more species slide ever closer to the brink.

One spring day in 1998 I stood atop the Darling escarpment, overlooking Perth. The countryside was lovely as only the west can be—wildflowers of astonishing variety bloomed everywhere, while the view down the scarp to the coastal plain, which had been greened by winter rain, was gentle and brimming with new life.

Despite the luxuriant prospect not all was well with the land. When it had been surveyed for wildlife a year or two earlier, the area proved almost devoid of marsupials. An extensive trapping program had located a single brushtail possum, which was captured no fewer than three times! But now a group had come together to breathe life back into this damaged landscape by dedicating close to 2000 hectares of private land as a new wildlife sanctuary. Known as Paruna, it was purchased by a non-profit organisation called the Australian Wildlife Conservancy (AWC), which is supported solely by the donations of ordinary Australians, who in this case had given six million dollars for the purchase.

Paruna links two national parks, neither of which at the time had much more wildlife than Paruna itself. Now, however, the whole region could be managed as a unit. Funds were raised to erect a fence that protected the reserve from farmland and houses, and a fox-baiting program was commenced in collaboration with the adjacent national parks. With fox numbers reduced, the first of the native mammals were released— ten black-flanked rock-wallabies (*Petrogale lateralis*).

By the time of their reintroduction to the region in 2000 these animals had been gone for over half a century, eliminated by hunting, land-clearance and introduced predators. No one was sure how they would cope, but within months they were breeding well, and today you can see them in many parts of Paruna and the adjoining national parks. They are an affirmation that our country can be healed, and that much of the work required for this to occur can be done by ordinary Australians.

The rock-wallabies were followed by introductions of short-nosed bandicoots, tammar wallabies and truffle-eating woylies (*Bettongia penicillata*), all of which have thrived and spread widely. Then in 2003 a small miracle happened—a long-vanished marsupial introduced itself to the reserve. The chuditch has become extinct over 95 per cent of its original distribution, with only a few remaining in the southwest of Western Australia. Its arrival at Paruna, unassisted by humans, indicates that the country has now begun to heal itself. Twelve months after the first chuditch sighting, thirteen more have appeared in the reserve, an indication that they are becoming re-established in country that has, for millions of years, been their home.

On 24 May 2004 I was once again in Western Australia on behalf of the AWC. This time I was gliding across the milky-emerald waters of Shark Bay, with Faure Island glowing ochre-red in the distance. The little-known island had been named in 1803 by French explorer Nicolas Baudin, who quickly moved on after one of his sailors was mauled by a tiger shark. The sharks could not long delay European settlement, however, and by the early twentieth century the island was a sheep run, with feral goats and cats compounding the damage done to the delicate desert ecosystem. By 2001 all of the island's native mammals—which had included woylies, western barred bandicoots (*Perameles bougainville*) and Shark Bay mice (*Pseudomys fieldi*)—had become extinct. The mouse, once common throughout the western two-thirds of the continent, by 1950 had been reduced to a single refuge—a few coastal sand-dunes on Bernier Island at the mouth of Shark Bay—making it among the most endangered of all Australia's mammals.

Faure Island is arid, and when I first saw it in 2001, its vegetation was as close-cropped as the stubble on a bikie's cranium. Just a handful of spindly, geriatric sandalwood trees had survived the onslaught of goats and sandalwood gatherers. Their nutritious fruit and spreading shade was thus no longer available to wildlife. Things began to change when Dick Hoult, fisherman, sheep farmer and scion of an extended family with roots in the Aboriginal community, passed the island's lease to the AWC. Dick and his family then became deeply involved in the restoration of Faure's environment. With the assistance of the Hoults and the Western Australian Department of Conservation, the cats, goats and sheep were removed, and improvements made to infrastructure. The day before we arrived was Dick's seventy-eighth birthday, which he celebrated by replacing the tin roof on Faure's main hut.

With his toes gripping the boat's gunnels like gnarled fingers, the old fisherman manoeuvred our vessel towards the shore. Faure had received thirty centimetres of rain in the first four months of 2004 and, in the absence of the sheep and goats, the island looked glorious. Delicate herbs and fragrant flowers covered the landscape, and even the mangroves surrounding its emerald-blue lagoons seemed revived by the rain. And once again the tracks of native mammals covered the island, for in 2003 the AWC had returned the boodie (*Bettongia lesueur*) and the Shark Bay mouse to their rightful home.

At first the reintroduction of the mice did not go smoothly, for a pair of boobook owls was nesting near their release site. The birds ate a dozen or so mice before moving on, but the survivors adjusted to the presence of the predators and have since bred apace, more than making up for the losses. Today, Faure is home to the largest population of Shark Bay mice in the world, and after decades of hovering on the brink, the species is on the road to recovery.

I was at Faure to release a third, very precious species in the new reserve: the banded hare wallaby (*Lagostrophus fasciatus*)—a relative of the short-faced kangaroos of the ice age—which by early 2004 was also facing extinction. The only wild populations inhabited Bernier and Dorre islands in Shark Bay, and a park ranger who had visited Bernier earlier in the year discovered that the few wallabies there were starving. Something had occurred on Bernier—and perhaps Dorre as well—that favoured other native species over the banded hare wallaby, and now, with only a handful remaining, it was likely that this wondrous creature was to become extinct.

With action urgently needed, national parks staff established a small captive group in an enclosure on the mainland. Six of these precious animals were to be released on Faure. Despite the fact that I had studied their fossil relatives for decades, I had seen nothing of the creatures but museum specimens, and so was unprepared for the magic of the living animal. They have large, dark and liquid eyes that are surrounded by white fur, and their muzzles give them something of the appearance of a lemur. They are soft and gentle creatures—even their footpads are remarkably soft—and their fur is a glorious greyish silver, banded in silver and dark chocolate on the back.

As the last rays of the sun fled the western sky a small group of AWC supporters stood amid the mulga, hessian sacks in hand. One by one we opened our sacks, releasing the banded hare wallabies. Most

Releasing one of seven banded hare wallabies on Faure Island in May 2004.

scampered off at once into the night, but the wallaby carried by the founder of the AWC, Martin Copley, wanted to stick about. It fed calmly, experimentally nibbling at the bushes a metre or so from us.

Half a dozen prior attempts have been made to establish colonies of banded hare wallabies, but all have failed; some because of cats, some because of eagles, and others from causes uncertain. As the last banded hare wallaby hopped calmly off into the night, I reflected on the nature of the age we live in. Those wallabies and their ancestors have been a part of my country for over 10 million years, but now, without human assistance, they might not even see out another decade. A responsibility comes with that knowledge, for with a little care they could enjoy a further 10 million years' tenure on our continent.

Australia was once dominated by people who loved the mother country—a land of lush greens and as alien to my country as any could be. Today Australians are more likely to proclaim a love of things native, yet because they often lack a true understanding of their environment, theirs is a love that can kill. Such well-meaning but uncomprehending enthusiasm is one reason why many Aboriginal communities continue to struggle under insupportable burdens, why native species keep vanishing, and why our future is being cut short by an insatiable addiction to fossil fuels. It is also why I wrote this book. We have now embarked on a new phase of our national existence, and just where it will lead I do not know. But I have a sinking feeling that unless every Australian searches profoundly for ways to help our land survive, things are likely to end badly for both ourselves and this great island continent.

Postscript

For the month following the release of the banded hare wallabies on Faure, AWC staff monitored them daily. The creatures quickly spread to the island's four corners, and at the end of this critical settling-in period they were all still alive. With each day that passed we felt more certain that the translocation would succeed. By the time spring brings its brief blossoming to the island there will be nine banded hare wallabies roaming Faure, for the three females released all had tiny, pink young in their pouches.

25 MILLION
YEARS AGO

10 MILLION
YEARS AGO

BALBARINES *(Balbarinae)*

BULUNGAMAYINES *(Bulungamayinae)*

THE FIRST
KANGAROO

THE KANGAROO FAMILY TREE

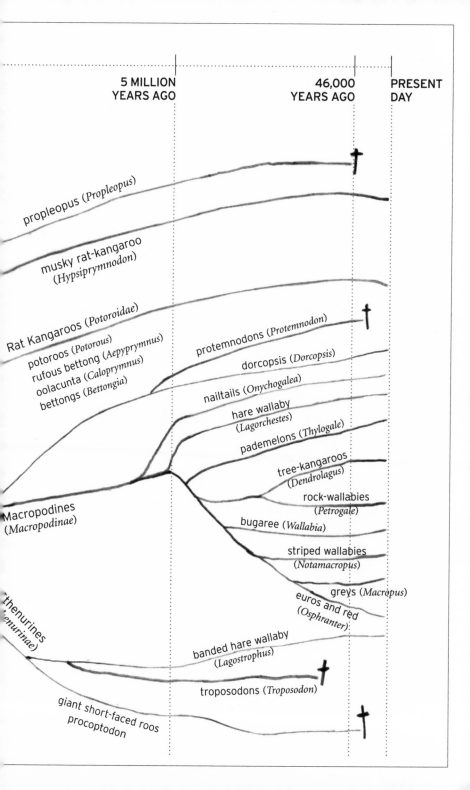

Acknowledgments

I owe Doctors Tom and Patricia Rich special thanks, for it was they who opened up a career in science to me, and who encouraged me to study kangaroos. As I was drafting this book Tom also provided details of our very early field work, before I kept field diaries.

Many people who have doubtless long forgotten me—Stolarski of Perth, the inhabitants of the Broome caravan park at Christmas 1975, and a huge number of country people Aboriginal and white—have extended exceptional generosity and hospitality to me as I have travelled the outback. They are the salt of the earth—the true Australians—and I thank them from the bottom of my heart for their many kindnesses. Bill Ellis, my long-suffering travelling companion, deserves special mention, and apologies, Bill, for abuse of the billy.

Australian science is, on the whole, warmly collegial. I thank those who have shared their discoveries with me and who have helped keep petty jealousies and destructive criticism out of our profession. Ron Strahan has been a lifelong friend and mentor, and I owe him particular recognition for several important historical works of his dealing with kangaroos.

As usual, Michael Heyward and Melanie Ostell have brought their exceptional editorial skills to the shaping of this work. And thank you to my family, who put up with my solitary confinement on weekends and evenings, as I wrote. Your support means everything to me.

General Bibliography

For those interested in the larger kangaroos Terry Dawson's excellent book *Kangaroos: Biology of the Largest Marsupials*, University of New South Wales Press, Sydney, 1995, is a must. An outstanding account of European perceptions of kangaroos (which also includes many images) is Ronald Younger's *Kangaroo: Images through the Ages*, Hutchinson, Sydney, 1988. *The Kangaroo Keepers*, Hugh Lavery ed., published by the University of Queensland Press, St Lucia, 1985, is an invaluable resource on kangaroos and kangaroo harvesting in Queensland. It includes an account of the saving of the bridled nailtail wallaby. A species by species description of the living kangaroos is included in *The Complete Book of Australian Mammals*, Ron Strahan ed., Cornstalk Publishing, Sydney, 1991, while an account of the extinct species can be found in *Prehistoric Mammals of Australia and New Guinea*, by John Long et al, University of New South Wales Press, Sydney, 2002. An important summary of studies relating to kangaroos (which also gives an indication of how the creatures were perceived at the time) is *Kangaroos and Men*, a special edition of the *Australian Zoologist*, vol. 16, part 1, 1971.

Historical works consulted include: Watkin Tench's account of early Sydney, republished as *1788*, edited and introduced by myself, Text Publishing, Melbourne, 1996; Ray Parkin's *H. M. Endeavour*, Miegunyah Press, Melbourne, 1997; H. H. Finlayson's classic *The Red Centre*, Angus & Robertson, Sydney, 1943, 4th edn; *A Truly Remarkable Man: The Life of H. H. Finlayson, and His Adventures in Central Australia*, Seaview Press, Henley Beach, 2001; Obed West's reminiscences, first published in the

Sydney Morning Herald in 1882, and reproduced as *The Memoirs of Obed West: A Portrait of Early Sydney*, Edward West Mariott ed., Barcom Press, Bowral, 1988, and *The Birth of Sydney*, Text Publishing, Melbourne, 1999. References to what the Dutch made of the quokka can be found in *Voyage of Discovery to Terra Australis, by Willem de Vlamingh in 1696–97*, Phillip Playford ed., Western Australian Museum, Perth, 1998.

References relating to ice-age kangaroos include my research on the diet of *Propleopus*, published in M. Archer and T. Flannery, 'Revision of the Extinct Gigantic Rat-Kangaroos (*Potoroidae, Marsupialia*), with Description of a New Miocene Genus and Species and a New Pleistocene Species of Propleopus', *Journal of Palaeontology*, 59, 1985, pp. 1311–49; W. D. L. Ride et al, 'Towards a Biology of *Propleopus oscillans* (*Marsupialia: Propleopinae, Hypsiprymnodontidae*)', *Proceedings of the Linnean Society of New South Wales* 117, 1997, pp. 223–328. An account of the Cuddie Springs site is J. H. Field and J. R. Dodson, 'Late Pleistocene Megafauna and Archaeology from Cuddie Springs, South-Eastern Australia', *Proceedings of the Prehistoric Society*, 65, 1999, pp. 275–301; Robert Boyle's experiments with things that glow are related in John Emsley's *The Shocking History of Phosphorus*, Macmillan, London, 2000; dating the timing of megafaunal extinction in Australia (which also gives evidence of the mixing of sand grains at Cuddie) is given in R. G. Roberts et al, 'New Ages for the Last Australian Megafauna: Continent-wide Extinction about 46,000 years ago', *Science*, 292, 2001, pp. 1888–92.

Nick Evans and Rhys Jones' seminal paper on the Pama-Nyungan language is 'The Cradle of the Pama-Nyungans: Archaeological and Linguistic Speculations', chapter 22 in *Archaeology and Linguistics: Aboriginal Australia in Global Perspective*, Patrick McConvell and Nicholas Evans eds, ANU Press, Canberra, 1997.

Important papers documenting the existence of ancient placental